Dynamics in Microwave Chemistry

Kama Huang • Xiaoqing Yang • Huacheng Zhu

Dynamics in Microwave Chemistry

Springer

Kama Huang
College of Electronics and Information Engineering
Sichuan University
Chengdu, China

Xiaoqing Yang
College of Electronics and Information Engineering
Sichuan University
Chengdu, China

Huacheng Zhu
College of Electronics and Information Engineering
Sichuan University
Chengdu, China

ISBN 978-981-15-9654-4 ISBN 978-981-15-9655-1 (eBook)
https://doi.org/10.1007/978-981-15-9655-1

This Springer imprint is published by the registered company Springer Nature Singapore Pte Ltd.
The registered company address is: 152 Beach Road, #21-01/04 Gateway East, Singapore 189721, Singapore

Contents

Principal List of Symbols

$\vec{B}$	Magnetic induction intensity
α, β_0	Electromagnetic wave propagation (attenuation constant, phase constant)
c	Velocity of light
C	Dielectric capacitance
C_ρ	Specific heat capacity
$\vec{D}$	Electric displacement
ε_0	Permittivity of free space
ε	Complex permittivity, $\varepsilon^{'} - j\varepsilon^{''}$
ε_r	Relative permittivity, $\varepsilon_r^{'} - j\varepsilon_r^{''}$
$\vec{E}$	Electric field
α, β_0	Loss tangent
f	Frequency
f_0	Resonant frequency
$\vec{F}_i$	Force of the *i*th atom
$\vec{H}$	Magnetic field
θ	Phase of reflection coefficients
$\vec{I}$	Matrix
$\overline{J}$	Jacobian tensor
$\vec{J^e}$	Induced current
k	Rate constant
k_0	Wave number in vacuum
k_c	Thermal conductivity
K_B	Boltzmann constant
L	Inductance
λ	Wavelength
m	Mass
N_A	Avogadro constant

n_e	Electron density
$\widehat{\text{n}}$	Normal to the conductor surface
$\vec{P}$	Polarization field vector
p	Pressure
q	Elementary charge
R	Resistance
$\vec{r}$	Position vector
ρ	Density
σ	Electrical conductivity
$\lvert S11 \rvert$	Amplitude of reflection coefficients (calculated, measured)
T	Temperature
t	Time
μ	Permeability
μ_r	Relative permeability
V	Volume
v	Velocity
ω	Angular frequency
φ	Phase angle
χ	Polarizability
Ω	Solid angle
τ	Relaxation time

Chapter 1
Introduction

In 1952, Broida et al. used microwave-induced plasmas to determine the deuterium isotope content in hydrogen–deuterium mixed gas [1], which created the application of microwave chemistry. Early in 1967, Williams reported that microwave heating can accelerate some chemical reactions. However, at that time, people were not aware that this discovery forecasted a revolution in chemistry. In 1986, Gedye et al. found that for 4-cyanatophenyl ions, the SN2 nucleophilic substitution reaction rate can be enhanced 1240 times with an improved yield by microwave radiation [2]. Thus, chemical reactions accelerated and controlled by microwave were taken into account. Presently, microwaves are widely applied in various chemical domains from inorganic reactions to organic reactions, from the medicinal chemical industry to the food chemical industry, and from simple molecules to complex life processes. Bose et al. placed reactants in a beaker filled with an ice–water mixture to ensure a constant temperature [3], and then obtained different results from conventional heating at the same temperature. Bogdal et al. researched different organic synthesis experiments and observed that microwave heating gives different reaction rates than conventional heating [4]. Eycken et al. found that the Diels-Alder reaction time can be shortened from 2 days to 13 min by microwave heating [5], and the reaction speed is increased 220 times. Agrawal et al. found that different results are produced in the cavity at the maximum electric field and magnetic field during material sintering [6]. Barnhardt et al. found that many chemical reactions that cannot be performed under low-temperature conditions can be performed by microwave irradiation under the same temperature conditions [7]. Akkali et al. indicated that thawing by microwave heating is faster than that by conventional methods [8]. Chandrasekaran et al. proved through experiments that the microwave heating technology can significantly improve the melting speed of aluminum, copper, and other metals [9]. Kolodiazhnyi reduced the melting temperature of ceramics to 1320 °C by microwave heating during ceramic sintering, and significantly improved the quality factor of the ceramics and reduced the energy consumption [10].

With a wide range of applications of microwaves in various chemical branches, conceptions established in conventional microwave heating and microwave

K. Huang et al., *Dynamics in Microwave Chemistry*,
https://doi.org/10.1007/978-981-15-9655-1_1

chemistry by using a commercial microwave oven have been severely challenged. Firstly, it is found that microwaves can accelerate chemical reactions while inhibiting the reactions under certain conditions. Besides, microwaves can change the reaction process. It has recently been found that the effect of microwaves on a chemical reaction may include a "non-thermal effect" as well as a thermal effect [11]. Loupy reported that there is a non-thermal effect of microwaves on chemical reactions [12]. Dudley and Regina experimentally demonstrated the non-thermal effect of microwaves on organic synthesis [13]. Note that a chemical reaction is a non-equilibrium system where old substances are continuously consumed, while new substances are continuously generated and the various phase interfaces may cause random variation. All these reasons lead to a nonlinear response of the reaction system to microwaves. It is necessary to clarify the interaction between microwaves and a chemical reaction in order to design effective and reasonable microwave chemical reactors.

At the same time, microwave heating is widely used in industrial applications because of its selectivity, rapidity, and high efficiency. Lai et al. could significantly shorten the degradation time of phenol by using a microwave-enhanced nickel oxide catalytic system [14]. Li et al. researched the mechanism for enhanced deamination by microwave irradiation, and believed that microwave irradiation caused the hydrogen bond vibrations in ammonia water to weaken the hydrogen bond energy [15]. In addition, the rapid heating of solvents by microwaves promoted the deamination process. Gedam et al. researched the microwave irradiation treatment of municipal solid waste, and found that when the microwave power is between 450 and 850 W, waste can be decomposed into gases such as carbon dioxide, carbon monoxide, and methane [16]. Lin et al. [17] and Solyom et al. [18] researched the effects of microwave irradiation on dehydration, drying, and other properties of urban sludge. However, microwave heating has problems such as non-uniform heating, overheating, and thermal runaway [19], which greatly limits its application. To overcome these existing problems, we investigate microwave heating by moving elements through simulation algorithms, which provide a new idea for the development of microwave foods.

To further study the interaction mechanism between microwaves and matter, we study the plasma generated by the interaction between microwaves and gases. We propose a multiphysics model to simulate the dynamic process of microwave plasma (from initial to equilibrium). Besides, the formation of a surface wave is calculated. Based on the multiphysics computation, some microwave plasma devices are designed for industrial applications.

The main goal of "Dynamics in Microwave Chemistry" is to elaborate the very complex scientific issues that pertain to the fundamental theory of microwave chemistry and the industrialized application in the near future. The book has seven chapters, which describe the theory and application of dynamics in microwave chemistry.

Chapter 2 introduces the dielectric properties in materials and in chemical reactions, as well as some methods of dielectric measurement under different conditions.

Chapter 3 introduces the dynamic analysis of microwave heating of chemical reactions by multiphysics computation, some algorithms, and multiphysics simulation.

Chapter 4 introduces two special phenomena of the interaction between microwaves and chemicals found in experiments.

Chapter 5 introduces some industrial applications in the field of microwave chemistry.

We provide specialist insight into the characterization and measurement of dielectric properties. Much attention is given to the theory of dynamics in microwave chemistry, including the effect of microwaves on the interactions between molecules and the algorithms and dynamic analysis of microwave heating and microwave plasma.

References

1. Nuechter M, Ondruschka B, Bonrath W, et al. Microwave-assisted synthesis: a critical technology overview. Chem Inform. 2004;6(3):128–41.
2. Gedye RN, et al. Microwave assisted syntheses in household microwave ovens. Tetrahedron Lett. 1986;27(3):279–83.
3. Bose AK, Manhas MS, Ghosh M, et al. ChemInform abstract: highly accelerated reactions in a microwave oven: synthesis of heterocycles. Chem Inform. 1990;21:47.
4. Bednarz S, Bogdal D. The comparative study of the kinetics of knoevenagel condensation under microwave and conventional conditions. The fifth international electronic conference on synthetic organic chemistry; 2001.
5. Van DEE, Appukkuttan P, De BW, et al. High-speed microwave-promoted hetero-diels-Alder reactions of 2(1H)-pyrazinones in ionic liquid doped solvents. J Org Chem. 2002;67 (22):7904–7.
6. Sato M, Roy R, Ramesh P, Agrawal D. Microscopic non-equilibrium heating a possible mechanism of microwave effects. International symposium on microwave science and its application to related fields, Japan; 2004. pp. 339–340.
7. Barnhardt EK. Advancing microwave energy to new heights with simultaneous cooling. European science foundation exploratory Workshop. Scientific report on microwave chemistry and specific microwave effects; 2005.
8. Akkari E, Chevallier S, Boillereaux L. Observer-based monitoring of thermal runaway in microwaves food defrosting. J Process Contr. 2006;16(9):993–1001.
9. Chandrasekaran S, Basak T, Ramanathan S. Experimental and theoretical investigation on microwave melting of metals. J Mater Process Technol. 2011;211(3):482–7.
10. Kolodiazhnyi T. $BaMg_{1/3}Nb_{2/3}O_3$-$Mg_4Nb_2O_9$ composite microwave ceramics with high Q-factor and low sintering temperature. J Eur Ceram Soc. 2012;32(16):4305–9.
11. Banik S, et al. Bioeffects of microwave—a brief review. Bioresour Technol. 2003;87(2):155–9.
12. Loupy A. Microwaves in organic synthesis. Weinheim: Wiley; 2004. p. 1–20.
13. Rosana MR, Tao Y, Stiegman AE, et al. On the rational design of microwave-actuated organic reactions. Chem Sci. 2012;3(4):1240–4.
14. Lai TL, Lee CC, Wu KS, et al. Microwave-enhanced catalytic degradation of phenol over nickel oxide. Appl Catal Environ. 2006;68(3–4):147–53.
15. Lin L, Yuan S, Chen J, et al. Removal of ammonia nitrogen in wastewater by microwave radiation. J Hazard Mater. 2009;168(2–3):862–7.

16. Gedam VV, Regupathi I. Pyrolysis of municipal solid waste for syngas production by microwave irradiation. Nat Resour Res. 2012;21(1):75–82.
17. Lin Q, Chen G, Liu Y. Scale-up of microwave heating process for the production of bio-oil from sewage sludge. J Anal Appl Pyrolysis. 2012;94:114–9.
18. Sólyom K, Mato RB, Pérez-Elvira SI, et al. The influence of the energy absorbed from microwave pretreatment on biogas production from secondary wastewater sludge. Bioresour Technol. 2011;102(23):10849–54.
19. Li ZY, Wang RF, Kudra T. Uniformity issue in microwave drying. Drying Technol. 2011;29 (6):652–60.

Chapter 2
Characterization and Measurement of a Chemical Reaction's Dielectric Properties

2.1 Dielectric Properties in Materials

In electromagnetic theory, the study of the interaction between an electric field and a dielectric needs to consider the complex permittivity ε. For general media, the complex permittivity is usually in plural form:

$$\varepsilon = \varepsilon' \text{-} j\varepsilon'' \tag{2.1}$$

where $\varepsilon^{'}$ is the real part of the complex permittivity and represents the ability to store electrical energy, $\varepsilon^{''}$ is the imaginary part of the complex permittivity and is considered as the heat loss in the dielectric due to the damping of the dipole vibration. In engineering, loss tangent is usually used to characterize the loss of the dielectric and the loss tangent is:

$$\tan\delta = \frac{\varepsilon''}{\varepsilon'} \tag{2.2}$$

which is seen to be the ratio of the real to the imaginary part of the total displacement current. Microwave materials are usually characterized by specifying the real part of the complex permittivity ($\varepsilon^{'}$) and the loss tangent at a certain frequency.

The complex permittivity represents the polarization and loss of the medium, which are closely related to the relaxation time and frequency.

2.1.1 Polarization

Dielectric material is a type of electrical insulator that can be polarized by an external electric field. In the presence of an electric field, the bound electric charges cannot

K. Huang et al., *Dynamics in Microwave Chemistry*,
https://doi.org/10.1007/978-981-15-9655-1_2

freely flow through the material, while they do slightly shift their average equilibrium positions, causing a dipole moment, namely dielectric polarization. The constitutive relation of matter in the presence of an external electric field can be expressed as

$$\vec{D} = \varepsilon_0 \vec{E} + \vec{P} \tag{2.3}$$

where $\vec{D}$ is the electric displacement vector, $\vec{E}$ is the intensity of the electric field, $\vec{P}$ is the dielectric polarization, and ε_0 is the permittivity of free space. According to the mechanism of polarization, dielectric polarization can be divided into three types: electronic polarization, ionic polarization, and orientation polarization.

A slight distortion of electron shells results in an induced dipole moment, and the whole process is called electronic polarization. The response time of the electrons to the electric field, which is called the relaxation time, is extremely short, e.g., 10^{-14} s ~ 10^{-16} s. At microwave frequency, the motion of the electrons will always be sufficiently rapid to build up a steady polarization. Hence, the electronic polarization can be expressed as

$$\overrightarrow{P_e} = N\alpha_e \vec{E} \tag{2.4}$$

where N is the total number of induced dipole moments per unit volume and α_e is the electronic polarizability.

Similarly, the deformation of nuclei positions also gives rise to an induced dipole moment, namely ionic polarization. The relaxation time of atomic polarization is also very short, e.g., 10^{-12} s ~ 10^{-13} s, which is in the infrared range. The atomic polarization in the steady state is determined by

$$\overrightarrow{P_a} = N\alpha_i \vec{E} \tag{2.5}$$

where α_i is the ionic polarizability.

The third important mechanism becomes effective and dominant if permanent dipole moments exist in the dielectric. In the presence of an applied electric field, the dipole moments tend to rotate along the direction of the electric field, giving rise to an over-all non-vanishing dipole moment, namely orientation polarization. Such a mechanism has a wide range of relaxation times, and the process is mainly determined by the structure of the dielectric. In the microwave frequency range, the motions of the dipole moments will not be sufficiently rapid to build up a time-dependent polarization that is in equilibrium with the electric field at any moment, leading to the collision of the dipole moments. This is the physical origin of the dielectric loss. The orientation polarization in the steady state is determined by

$$\overrightarrow{P_o} = N\alpha_o\vec{E} \tag{2.6}$$

where α_o is the orientation polarizability.

Considering the three types of polarization mechanism in Fig. 2.1, dielectric polarization can be expressed by

$$\vec{P} = \overrightarrow{P_e} + \vec{P_i} + \overrightarrow{P_o} \tag{2.7}$$

In nonpolar dielectrics, only electronic polarization and ionic polarization occur in the presence of an electric field. While all the three mechanisms will occur in a polar dielectric, orientation polarization plays a dominant role in determining the dielectric property of the dielectric. For media that are isotropic and linear, one can obtain

$$\vec{D} = \varepsilon_0\varepsilon_r\vec{E} = \varepsilon_0(\chi + 1)\vec{E} \tag{2.8}$$

where ε_r is the complex permittivity and χ is the polarizability.

2.1.2 *Complex Permittivity and Loss*

The process of converting electrical energy into heat energy under the action of an external electric field is called *dielectric loss*, which directly results in an increase in dielectric temperature. The main causes of dielectric loss include (1) conductance loss, (2) relaxation polarization loss, and 3) dispersion loss.

The dielectric is not an ideal insulator, so there must be some weakly associated conductive carriers. Under an electric field, the carriers drift to form a current, which is consumed in the form of heat, namely conductance loss.

The dielectric is polarized during the action of the electric field, and the establishment of various polarizations takes certain time. The time required for the establishment of thermal ion relaxation polarization and dipole steering polarization is relatively long, which is 10^{-2}–10^{-8} s, or even longer. When the external electric field frequency is low, such polarization can keep up with the cyclical change in the external electric field and the polarization can be completed. However, when the frequency of the external electric field is relatively high, such as high frequency or ultra-high frequency, the dipole turning polarization cannot keep up with the periodic variation in the external electric field, resulting in relaxation. This causes the polarization $\vec{P}$ of the dielectric to lag behind the external electric field strength $\vec{E}$, and as the frequency of the external electric field increases, the complex permittivity of the dielectric decreases. When the external electric field frequency is sufficiently high and the dipole steering polarization cannot keep up with the periodic variation of the electric field, the complex permittivity provided by this polarization form

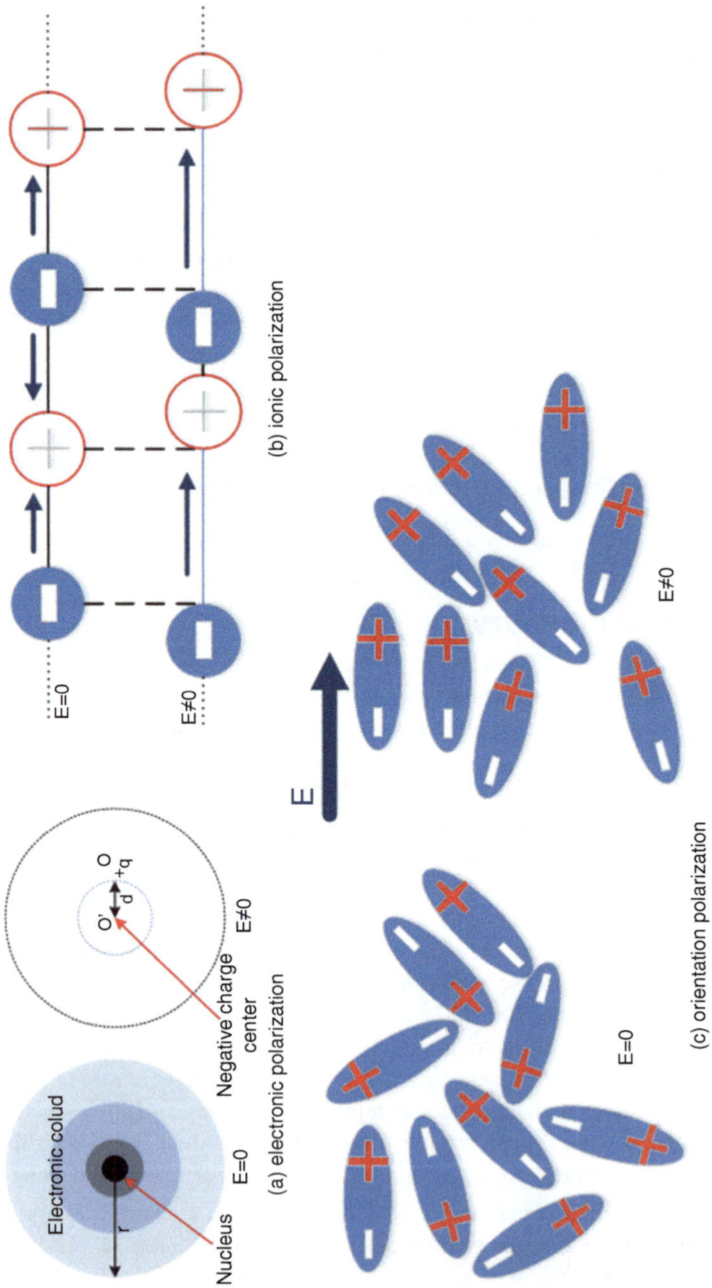

Fig. 2.1 Various forms of polarization

decreases to zero with increasing frequency, i.e., it tends to the optical frequency dielectric coefficient. This process also consumes energy, and at high or ultra-high frequencies, such losses will play a major role, even greater than the conductance loss. The above discussion concerns the relaxation polarization loss.

Dispersion loss is derived from the resonance effect of atoms, ions, or electrons when vibrating or rotating. This effect occurs in the infrared-to-ultraviolet range, and hence, will not be discussed in detail.

It can be seen from the above discussion that the complex permittivity, dielectric loss, and loss tangent are all related to frequency, specifically:

When $\omega \rightarrow 0$ (where ω is angular frequency), the complex permittivity ε reaches the maximum possible value, dielectric loss is mainly the conductance loss, and tanδ tends to infinity.

When $\omega\tau < <1$ (where τ is relaxation time), the relaxation polarization cannot be fully established, the complex permittivity ε will decrease, losses due to relaxation polarization begin to appear, and tanδ is inversely proportional to the frequency and sharply decreases.

When $\omega\tau = 1$, the relaxed polarization is most sensitive to changes in the electric field; thus, ε changes rapidly with frequency, the polarization loss increases significantly, and tanδ has a local maximum value.

When $\omega\tau > > 1$, the relaxation polarization is too slow to be established, so ε tends to optical permittivity($\varepsilon \rightarrow \varepsilon_\infty$), the loss tends to a fixed value, and tanδ tends to zero.

In summary, the dielectric properties of matter are closely related to the relaxation time and frequency of the electric field. In the next section, the time and frequency domains will be discussed in detail.

2.2 Dielectric Polarization in Chemical Reactions

2.2.1 Challenges to Characterize Dielectric Behaviors in Chemical Reactions

In Sect. 2.1, the orientation polarization mainly describes the states of polar molecules in the presence of an electric field. However, it alone cannot uniquely specify the states of polar-molecule reactions in the liquid phase, for the reactions are typically non-equilibrium systems and the components change during the chemical process. Each type of molecule has a specific dipole moment. Therefore, the amount and intensity of the overall dipole moments are related to the chemical process. Thus, the chemical process gives rise to a positive or negative polarization increment. In the presence of an applied field, the polarization of polar molecule reactions in the liquid phase should be physically determined by the orientation polarization of the polar molecules and polarization increment caused by the chemical process. To understand the polarization process of chemical reactions irradiated by microwaves,

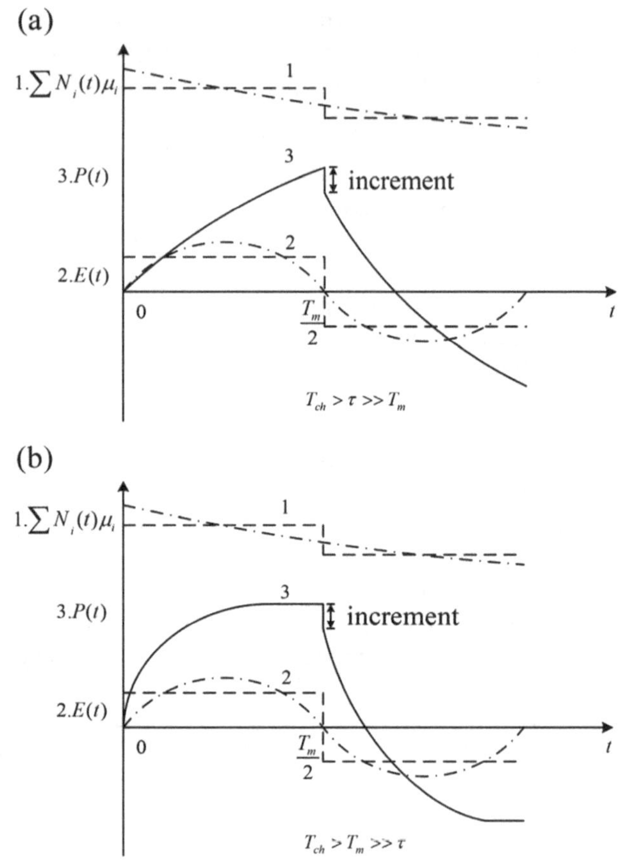

Fig. 2.2 Polarization of chemical reactions in the presence of a sinusoidal electric field

the relations among the reaction time (T_{ch}), relaxation time of the dipole rotation (τ), and microwave cycle (T_m) need to be established. There are six different cases among them. We are concerned with the assumption that T_{ch} is larger than τ and T_m. Such an assumption can be easily justified in organic chemistry.

Based on the common assumption of local equilibrium, the polarization of a chemical reaction in the presence of a sinusoidal electric field is shown in Fig. 2.2, where the subscript i represents the reactants. We suppose that the amount and intensity of the overall dipole moments in the chemical reaction decline during the chemical process. As shown by the dashed lines, we discretize the sinusoidal electric field into a series of square-wave electric fields and divide the chemical process into a series of equilibrium systems. The magnitude of the square-wave electric field is equal to the effective value of the sinusoidal electric field.

The polarization, shown as a solid line, is obviously nonexistent at the initial time. When an external electric field is applied, each dipole moment will tend to line up with the field and the polarization will increase with the field. In Fig. 2.2a, the polarization is always unsteady in the first half cycle of the square wave and the relaxation time of the dipole rotation is larger than the half cycle. However, there is also a contrary phenomenon in Fig. 2.2b. The polarization in both cases is the same as that in the liquid mixtures. However, when $t = T_m/2$, there will be a sudden negative increment in polarization because of the chemical process. Thus, the polarization of the chemical reaction cannot be described by orientation polarization.

2.2.2 Dielectric Polarization of Diluted Polar-Molecule Reactions

The general chemical reaction process can be expressed as

$$\sum_i X_i A_i \xrightarrow{k} \sum_j Y_j B_j \tag{2.9}$$

Where k is the rate constant, X and Y are the stoichiometric coefficients, i and j represent the reagents, and A and B denote the reactants and products, respectively. In the presence of an electric field, the polar molecules will not only be rotating with respect to the direction of the applied field but also splitting and combining stochastically. Thus, in polar-molecule reactions, the polarization will depend on the rotational Brownian motion and the chemical process. According to the continuity equation, the probability distribution function of the molecular concentration can be determined by the modified Smoluchowski–Debye equation:

$$\frac{\partial \gamma_{A_i}(\theta,t)}{\partial t} = -v + \frac{D_{\gamma_{A_i}}}{\sin\theta}\frac{\partial}{\partial\theta}\left\{\sin\theta\left[\frac{\partial \gamma_{A_i}(\theta,t)}{\partial\theta} + \frac{1}{k_B T}\gamma_{A_i}(\theta,t)\frac{\partial U_{A_i}}{\partial\theta}\right]\right\} \tag{2.10}$$

$$\frac{\partial \gamma_{B_j}(\theta,t)}{\partial t} = v + \frac{D_{\gamma_{B_j}}}{\sin\theta}\frac{\partial}{\partial\theta}\left\{\sin\theta\left[\frac{\partial \gamma_{B_j}(\theta,t)}{\partial\theta} + \frac{1}{k_B T}\gamma_{B_j}(\theta,t)\frac{\partial U_{B_j}}{\partial\theta}\right]\right\} \tag{2.11}$$

where $v = -\left(\frac{\partial \gamma_{A_i}}{\partial t}\right)_{\text{chem}} = \left(\frac{\partial \gamma_{B_j}}{\partial t}\right)_{\text{chem}}$ represents the effect of the chemical reaction rate. In Eqs. (2.10) and (2.11), the rotational diffusion coefficients are assumed to be constant, as they only change slightly during a chemical process. The concentrations of the molecules are represented by

$$dc_{A_i}(t) = \gamma_{A_i}(\theta,t)d\Omega \tag{2.12}$$

$$dc_{B_j}(t) = \gamma_{B_j}(\theta,t)d\Omega \tag{2.13}$$

To solve Eqs. (2.10) and (2.11), an expression for v should be established. For easier presentation, we now consider solutions of A_1 and B_1. According to the chemical kinetics, we can get

$$\left(\frac{\partial c_{A_1}}{\partial t}\right)_{\text{chem}} = -k\prod_i c_{A_i}^{X_i} \tag{2.14}$$

Substituting Eq. (2.12) into Eq. (2.14), we find

$$\left(\frac{\partial \gamma_{A_i}}{\partial t}\right)_{\text{chem}} = -\frac{k\gamma_{A_1}\prod_i c_{A_i}^{X_i}}{c_{A_1}} \tag{2.15}$$

Now, the modified Smoluchowski equation becomes similar to the diffusion-reaction equation. For polar molecules with a permanent dipole moment, the distribution function can be expressed by the standard basis of Legendre polynomials:

$$\gamma_{A_1}(\theta, t) = \sum_{n=0}^{\infty} \frac{2n+1}{2} a_{1,n}(t) P_n(\cos\theta) \tag{2.16}$$

$$\gamma_{B_1}(\theta, t) = \sum_{n=0}^{\infty} \frac{2n+1}{2} b_{1,n}(t) P_n(\cos\theta) \tag{2.17}$$

where $a_{1,\ n}(t)$ and $b_{1,\ n}(t)$ are the corresponding order functions. By combining Eqs. (2.10), (2.15), and (2.16) and using the orthogonality of the Legendre polynomials, we obtain

$$\begin{aligned}\frac{\partial a_{1,n}}{\partial t} = & -\frac{k\prod_i c_{A_i}^{X_i}}{c_{A_1}} a_{1,n} - n(n+1)D_{r_{A_1}} a_{1,n} \\ & + u_{A_1}\frac{n(n+1)}{2n+1} D_{r_{A_1}}(a_{1,n-1} - a_{1,n+1})\end{aligned} \tag{2.18}$$

Similarly, we can get

$$\begin{aligned}\frac{\partial b_{1,n}}{\partial t} = & -\frac{k\prod_i c_{A_i}^{X_i}}{c_{A_1}} b_{1,n} - n(n+1)D_{r_{A_1}} b_{1,n} \\ & + u_{B_1}\frac{n(n+1)}{2n+1} D_{r_{B_1}}(b_{1,n-1} - b_{1,n+1})\end{aligned} \tag{2.19}$$

It seems that it is very challenging to solve Eqs. (2.18) and (2.19). Fortunately, according to Eq. (2.13) and the orthogonality of the Legendre polynomials, we can obtain the polarization

$$P = \sum_i \frac{\mu_{A_i} N_A c_{A_i}(t) a_{i,1}(t)}{b_{i,0}(t)} + \sum_j \frac{\mu_{B_j} N_A c_{B_j}(t) b_{j,1}(t)}{b_{j,0}(t)} \tag{2.20}$$

where N_A is Avogadro constant. It is now obvious that we only need to obtain the expressions of $a_{1,0}$, $a_{1,1}$, $b_{1,0}$, and $b_{1,1}$.

Setting $n = 0$, we can easily find

$$\frac{\partial a_{1,0}}{\partial t} = -\frac{k\prod_i c_{A_i}^{X_i}}{c_{A_1}} a_{1,0} \tag{2.21}$$

$$\frac{\partial b_{1,0}}{\partial t} = -\frac{k\prod_i c_{A_i}^{X_i}}{c_{A_1}} b_{1,0} \tag{2.22}$$

The solution of Eq. (2.21) is obviously

$$a_{1,0}(t) = c_{A_1}(t) \tag{2.23}$$

Thus, we can easily obtain

$$b_{1,0}(t) = c_{B_1}(t) \tag{2.24}$$

By neglecting the high order of the Legendre polynomials, we can now obtain $a_{1,1}$ and $b_{1,1}$:

$$\begin{aligned} a_{1,1} = {} & \frac{2\mu_{A_1} D_{r_{A_1}}}{3k_B T} \exp\left[-\int\left(k\prod_i c_{A_i}^{X_i}(t)/c_{A_1}(t) + 2D_{r_{A_1}}\right)dt\right] \\ & \times \int_0^t \frac{1}{\exp\left[-\int\left(k\prod_i c_{A_i}^{X_i}(t')/c_{A_1}(t') + 2D_{r_{A_1}}\right)dt'\right]} a_{1,0}(t')E(t')dt' \end{aligned} \tag{2.25}$$

$$\begin{aligned} b_{1,1} = {} & \exp\left(-2D_{r_{B_1}} t\right) \int_0^t \frac{1}{\exp\left(-2D_{r_{B_1}} t'\right)} \\ & \times \left[A_{1,1}(t')k\prod_i c_{A_i}^{X_i}(t')/c_{A_i}(t')\right] + \frac{2\mu_{B_1} D_{r_{B_1}}}{3k_B T} b_{1,0}(t')E(t')dt' \end{aligned} \tag{2.26}$$

The solutions for the other reactants and products can be achieved similarly, and the final expression of polarization depends on the first-order solution of the Legendre polynomials:

$$P = \sum_i \mu_{A_i} N_A a_{i,1}(t) + \sum_j \mu_{B_j} N_A b_{j,1}(t) \tag{2.27}$$

This study aims to theoretically investigate the dielectric polarization in polar-molecule reactions in the liquid phase. On the basis of the modified Smoluchowski equation, the analytic expression of polarization is derived. The result shows that the expression of polarization in polar-molecule reactions is quite different from that in the Debye model. For general reactions, two vectors, the rotational diffusion vector and the component concentration vector, are proposed to describe the polarization. For fast reactions, the concept of permittivity will be nonexistent due to the effect of the component concentration vector on the frequency of the applied field and the dielectric behavior may only be analyzed in the time domain. On the contrary, for slow reactions, the permittivity will be concentration-dependent.

2.2.3 *Simplified Expression of Polarization in Diluted Polar-Molecule Reactions*

Equation (2.27) shows that the polarization in polar-molecule reactions is quite different from that in polar molecules in the time domain, and that it is not easy to simplify Eq. (2.27) due to the complex mathematical integral. Fortunately, the relaxation time of a dipole moment is always very smallin practice, especially so for small molecular dipoles, about 10^{-11} s. That is, the relaxation time of the dipole moment is much smaller than the chemical reaction time, which leads to

$$k \prod_i c_{A_i}^{X_i}(t) / c_{A_1}(t) << 2 D_{r_{A_1}} \tag{2.28}$$

Now, Eqs. (2.24) and (2.25) become

$$a_{1,1} = \frac{2\mu_{A_1} D_{r_{A_1}}}{3 k_B T} e^{-\frac{t}{\tau_{A_1}}} * [a_{1,0}(t) E(t)] \tag{2.29}$$

$$b_{1,1} = k \prod_i c_{A_i}^{X_i}(t) / c_{A_i}(t) \cdot e^{-\frac{t}{\tau_{B_1}}} * b_{1,1}(t) + \frac{2\mu_{B_1} D_{r_{B_1}}}{3 k_B T} e^{-\frac{t}{\tau_{B_1}}} * [b_{1,0}(t) E(t)] \tag{2.30}$$

In most cases, the first term on the right side of Eq. (2.30) is much smaller than that of Eq. (2.29). Hence, by combining Eqs. (2.21, 2.22, 2.27, 2.29, and 2.30), we can obtain a simplified expression of polarization:

$$\begin{aligned}\vec{P}(t) &= \sum_i \phi_{A_i}(t) * \left[c_{A_i}(t)\vec{E}(t)\right] + \sum_j \phi_{B_j}(t) * \left[c_{B_j}(t)\vec{E}(t)\right] \\ &= \widehat{\varphi}(t) * \left[\widehat{C}(t)\vec{E}(t)\right]\end{aligned} \tag{2.31}$$

Where $\varphi(t)$ is a function that contains the rotational diffusion property of the dipole moment. Here, we define a new type of convolution for the vectors, where

$$\widehat{\phi}(t) = \left[\varphi_{A_1}(t), \varphi_{A_2}(t), \ldots, \varphi_{A_i}(t), \varphi_{B_1}(t), \varphi_{B_1}(t), \ldots, \varphi_{B_j}(t)\right] \tag{2.32}$$

is the rotational diffusion vector and

$$\widehat{C}(t) = \left[c_{A_1}(t), c_{A_2}(t), \ldots, c_{A_i}(t), c_{B_1}(t), c_{B_1}(t), \ldots, c_{B_j}(t)\right]' \tag{2.33}$$

is the component concentration vector.

We can easily see that the main difference between Eqs. (2.31) and (2.20) is the extra term $\widehat{C}(t)$. In other words, $f(t)$ contains the rotational property of the dipole moments. However, in Eq. (2.31), $\widehat{\phi}(t)$ contains the rotational property of the dipole moments, while $\widehat{C}(t)$ represents the property of number. We can now use the Fourier transform, and the polarization in polar molecules and polar-molecule reactions in the frequency domain can be expressed as

$$\vec{P}_p(\omega) = f(\omega)\vec{E}(\omega), \tag{2.34}$$

$$\vec{P}_r(\omega) = \frac{1}{2\pi}\widehat{\phi}(\omega)\left[\widehat{C}(\omega) * \vec{E}(\omega)\right]. \tag{2.35}$$

For polar molecules, the polarization is mainly the multiplication of the response function $f(\omega)$ and the applied field $\vec{E}(\omega)$. The frequency of the electric field will not be affected by the response function, and the relative complex permittivity is calculated from this formula. On the other hand, for a polar-molecular reaction, the convolution leads to the frequency variation of the electric field. For fast reactions, in which the chemical reaction rate is comparable to, or larger than, the rate of change of the electric field, there will no longer be a relative permittivity; the dielectric behavior may only be analyzed in the time domain. For slow reactions, the chemical reaction rate is almost constant compared to the rate of change of the electric field. Equation (2.35) becomes

$$\vec{P}_r(\omega) = \widehat{C}'\widehat{\phi}'(\omega)\vec{E}(\omega). \tag{2.36}$$

In this case, the permittivity can be derived from Eq. (2.36), which is similar to Eq. (2.34). Note that the relative permittivity is concentration-dependent, for the concentration slowly changes with time.

2.3 Relative Permittivity Measurement

2.3.1 Typical Dielectric Measurement Methods

The measurement of relative dielectric properties at microwave and millimeter-wave frequencies has gained increasing importance, especially in research fields such as material science, microwave circuit design, absorber development, and biological research. Dielectric measurement is important because it can provide the electrical or magnetic characteristics of the materials, which prove useful in many research and development fields.

There are many techniques available for measuring the dielectric properties of materials. Each method has its own strengths and limitations for a particular application, which depend on frequencies, material properties, temperature, destructive or non-destructive nature, contacting or non-contacting nature, etc. Here are a few traditional dielectric measurement methods that we will discuss:

2.3.1.1 Transmission/Reflection Line Method

The transmission/reflection line method is a popular broadband measurement method that involves placing a material under test (MUT) in a section of a waveguide or coaxial line and measuring the two-port complex scattering parameters (S11, S21) using a vector network analyzer (VNA), as seen in Fig. 2.3. The conversion of the S-parameters to a complex dielectric parameter is computed by solving the equations using a program. This method can measure the transformation of the transmission- and reflection-wave minimum points, and avoid the iterative solution of a complex transcendental equation. However, this method is limited to low- and medium-loss media. For high-loss media, there are no multiple reflections in the sample. The transfer line method is applicable to solids and liquids with large permittivity, but is not suitable for a gas with low permittivity.

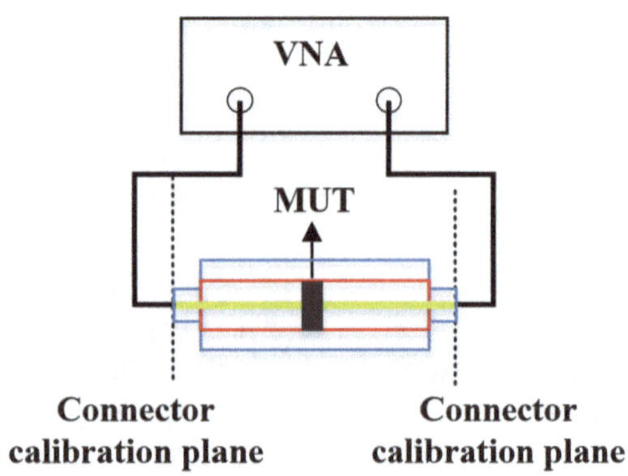

Fig. 2.3 Transmission/reflection line method

2.3.1.2 Open-Ended Coaxial Probe Method

The open-ended coaxial probe method is a non-destructive testing method. Here, the tip of the probe is immersed into the liquid sample under test, or can be pressed against a solid sample, and then, S11 or the reflection coefficient is measured using a network analyzer, as seen in Fig. 2.4. This method is mature for measuring the relative permittivitys of liquid and semi-fluid media. However, there are still some areas that need to be improved for measuring the relative permittivitys of solid materials, especially hard and non-planar solid materials.

2.3.1.3 Free-Space Method

Free-space measurement is suitable for hostile environments and generally operates at wide-band frequencies, comprising a vector network analyzer and two antennae facing each other with a sample holder between them for obtaining S11 and S21. The measurement requires the MUT to be large and flat, as seen in Fig. 2.5. In addition, the free-space method requires the material to have high loss, or else it will generate standing waves and cause errors. Therefore, this method is only applicable to high frequencies above 3 GHz, reaching up to 100 GHz.

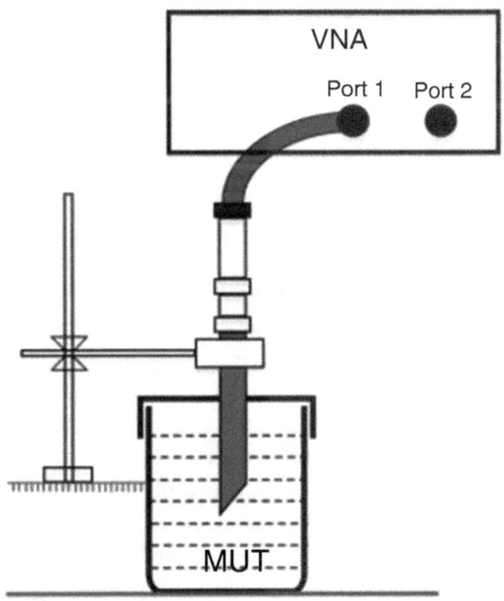

Fig. 2.4 Open-ended coaxial probe method

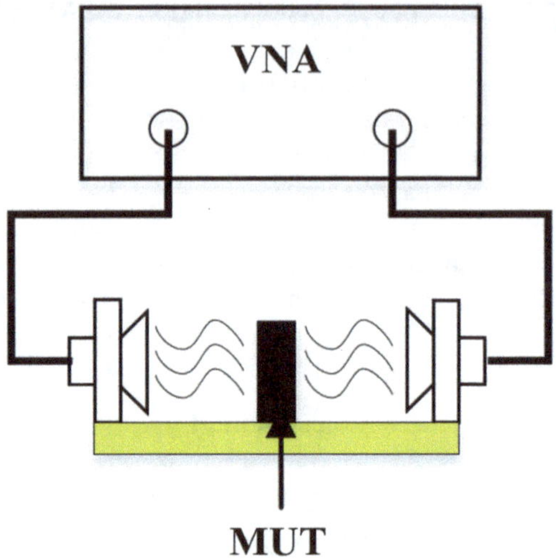

Fig. 2.5 Free-space method

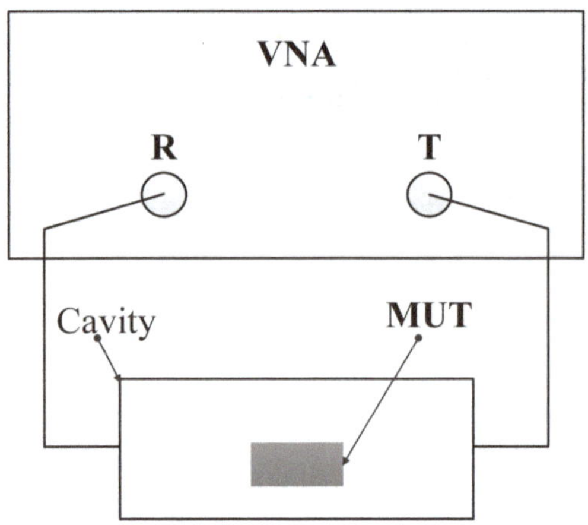

Fig. 2.6 Resonant method

2.3.1.4 Resonant Method

The resonant method shown in Fig. 2.6 is used to measure the relative permittivity of the sample as part of the resonant structure, which is based on the disturbance theory. The method requires small amounts of a liquid sample and exhibits advantages of high sensitivity and accuracy. Notably, the measurement of resonant frequencies and the Q-factor of a cavity can give satisfactory results. The resonance method can measure the relative permittivity of all materials within the frequency range.

Table 2.1 describes some of the measured conditions, parameters, advantages, and disadvantages of the various relative permittivity measurement methods.

Note that the relative permittivity measurements of materials are important in microwave engineering and microwave chemistry so that they can be used to

Table 2.1 Comparison of relative permittivity measurement methods

Measurement techniques	Loss	Measured parameters	Advantages	Disadvantages
Transmission/ reflection line method	Medium to high loss	S11, S21	Broadband measurement	Low accuracy
Open ended coaxial probe method	Lossy	S11	Broadband, easy sample preparation	Multi-valued problem
Free space method	Medium loss	S11, S21	Broadband, non-destructive	Large and flat MUT, Diffraction effects
Resonant cavity method	Low loss	Frequencies, Q-factors	High accuracy	Need high frequency resolution VNA, narrow band

describe the transmission and absorption of microwaves in a mixture during a chemical reaction. However, there are some disadvantages of the relative permittivity measurements, such as calibration, a small dynamic measurement range, strict measurement size requirements, or when being used in in some special scenes. In addition, during the analysis, electromagnetic transmission is not as complicated as the analytical expressions or analytical solutions.

Nowadays, with the development of the microwave energy industry, traditional dielectric measurement methods are habitual but unadaptive. The relative permittivity of materials is not stable, which means it changes with microwave frequency, temperature, and reaction time. For the reactants from a complicated mixture, which varies with time, an relative permittivity can be used to describe the molecular polarization of the mixture in the reaction. The relative permittivity is expected to vary with respect to the microwave frequency, temperature, and reaction time.

To overcome the difficulties that arise in the application of microwave energy in a reaction system, the relative permittivity of a mixture during a chemical reaction needs to be carefully studied. So far, some efficient numerical approaches have been used to study the microwave heating of water and food. In many cases, the relative permittivity of the chemical reaction is difficult to be measured online using traditional reconstruction algorithms. Therefore, novel relative permittivity measurement apparatus with modern optimal algorithms are in demand for chemical reaction systems.

This section introduces several methods that use transmission-only scattering (S-) parameter measurements to quickly and accurately characterize the dielectric properties by a genetic inverse algorithm and neural network reconstruction algorithms. The measured examples are given to verify their correctness and reliability in microwave chemistry. Finally, the dielectric measurements under different conditions are listed based on our recent research into reaction systems.

2.3.2 *Modern Optimal Algorithm in Dielectric Measurements*

In the microwave chemistry research, it is very important to measure the equivalent relative permittivity accurately. Section 3.1 introduces some analytical methods for the relative permittivity measurement. These analytical methods consider some complicated electromagnetic field boundary problems, which are not suitable for some medium- and high-dielectric-loss media. Numerical calculation is used to measure the relative permittivity and there are some errors because it is difficult to simulate the experimental device accurately.

Due to the limitations of the above methods, we propose to use some algorithms to measure the relative permittivity, such as a genetic algorithm or neural network. In the following sections, we will introduce the characteristics and application scenarios of these measurement algorithms.

2.3.2.1 Genetic Algorithm

Introduction

The inverse problem of an electromagnetic field is the determination of the ideal performance parameters or indicators of electromagnetic field devices. To reach this goal, it is necessary to optimize the device design. In the process of design optimization, the numerical computation of the electromagnetic field must be involved. Genetic algorithm (GA) is a type of numerical computation method, proposed by Holland in 1975 [1]. As a new global optimization algorithm, it is widely used, robust, and suitable for parallel processing. In dielectric measurements, a GA is mainly used to obtain the optimal solution of the relative permittivity of the material through the optimization algorithm when there is no analytical solution.

In the following sections, we will introduce some implementations and applications of a GA in dielectric measurements.

Application of GA

This section introduces a hybrid experimental/computational method for determining the "relative permittivity" in materials and reaction systems.

A new resonant structure of a coaxial line that measures the reflection coefficients is designed. The slotted open-ended coaxial line (Fig. 2.7) works as a cavity, which is loaded with the solution under test, to measure the reflection coefficient, which is sensitive to small changes in the relative permittivity of the solution in the cavity. The slot is made for pushing air bubbles out of the cavity when the sensor is immersed in the solution.

To predict the performance of the resonant coaxial structure, the electromagnetic-field distribution near the sensor and the reflection coefficients at different

Fig. 2.7 Coaxial sensor

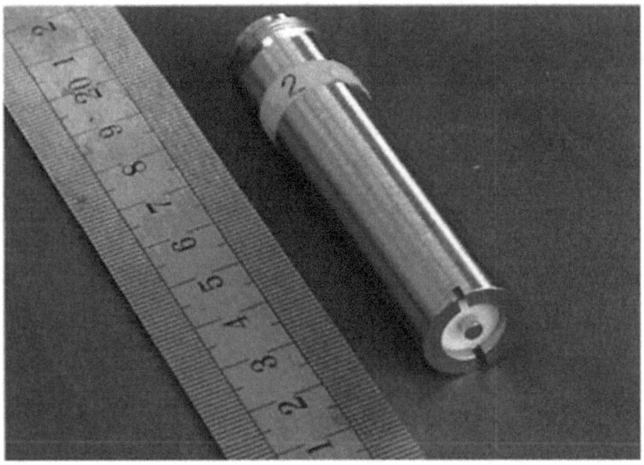

frequencies are calculated. In the calculation, a frequency-dependent finite-difference time-domain (FDTD) method is employed due to the dilute solution that is to be studied in a dispersive material.

Then a GA-based inverse-calculation technique is developed and employed to determine the complex permittivity of pure water from the measured reflection coefficient and the results are compared with those obtained from Debye's equation.

A block diagram of the GA-based inverse calculation is illustrated in Fig. 2.8. As the first step of the inverse calculation, a set of reflection coefficients is obtained by forward calculation employing the FDTD method from a randomly initiated relative permittivity. The set of relative permittivity constructs, the initial population of the potential solutions, and the reflection coefficients are transformed into objective values by using the objective function, defined by

$$f(x_i) = \left[\alpha_1(|S11_c| - |S11_m|)^2 + \alpha_2(\theta_c - \theta_m)\right]^{1/2} \tag{2.37}$$

where $|S11_c|$ and θ_c are the magnitude and phase of a set of the calculated reflection coefficients, respectively; $|S11_m|$ and θ_m are the amplitude and phase of the measured reflection coefficients, and α_i $(i = 1, 2)$ is the weight factor. In this computation, $\alpha_1 = 0.9$ and $\alpha_2 = 0.15$. The objective values are then mapped onto the fitness values by using the fitness function defined by

$$F = \exp\left[-\alpha_0 f(x_i)\right] \tag{2.38}$$

where α_0 is an empirical parameter, and is taken as 0.33 in our calculation. To select better individuals in the population, some special techniques, such as stochastic universal sampling (SUS) and the elitist strategy are employed. The evolution of the survival of the fitness terminates under the condition that the fitness value is larger than a preset threshold F_0.

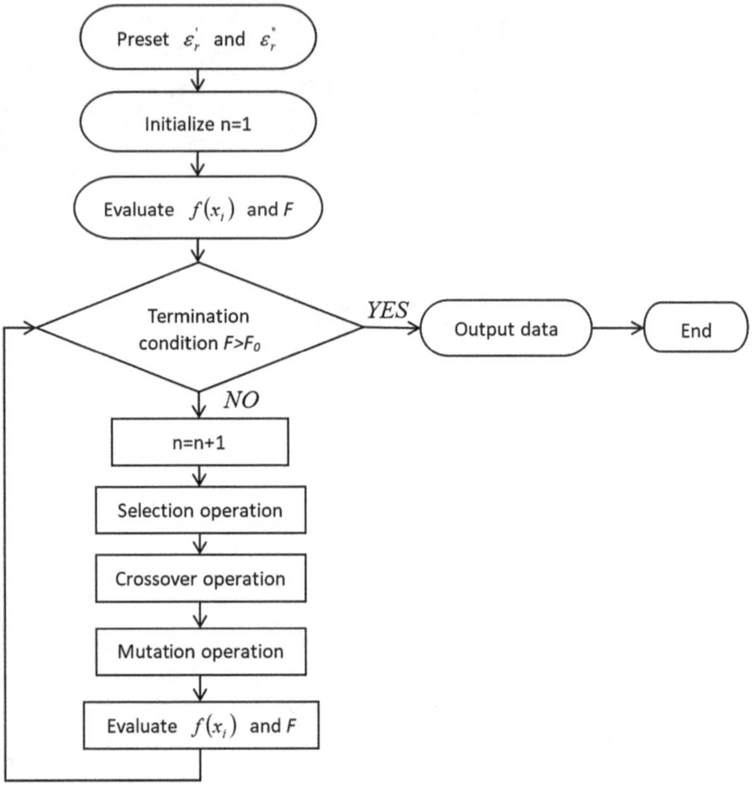

Fig. 2.8 Inverse-calculation procedure

Table 2.2 Comparison of the relative permittivity obtained by GA with that by Debye's equation in pure water

	40°C		50°C		60°C		70°C		80°C	
	ε^*_{eff}	ε^*_{eff}	ε^*_{eff}	ε^*_{eff}	ε^*_{eff}	ε^*_{eff}	ε^*_{eff}	ε^*_{eff}	ε^*_{eff}	ε^*_{eff}
GA result	70.72	5.63	69.02	4.85	66.07	3.37	61.87	2.97	59.38	2.22
Debye's result	71.63	6.02	68.34	4.63	65.28	3.61	62.44	2.85	59.82	2.28
Relative errors	–1.3%	–6.9%	1.0%	4.5%	1.2%	–7.1%	–0.9%	4.0%	–0.7%	–2.7%

Using the measurement apparatus presented above, the reflection coefficient of pure water can be measured. From the measured reflection coefficient, the relative permittivity of the solution is then calculated by employing the GA-based inverse-calculation method described in Fig. 2.8. The computational results of the relative permittivity of pure water determined by GA optimization are compared with those obtained from Debye's equation at a frequency of 2450 MHz. The comparison is listed in Table 2.2, and a reasonably good agreement is observed.

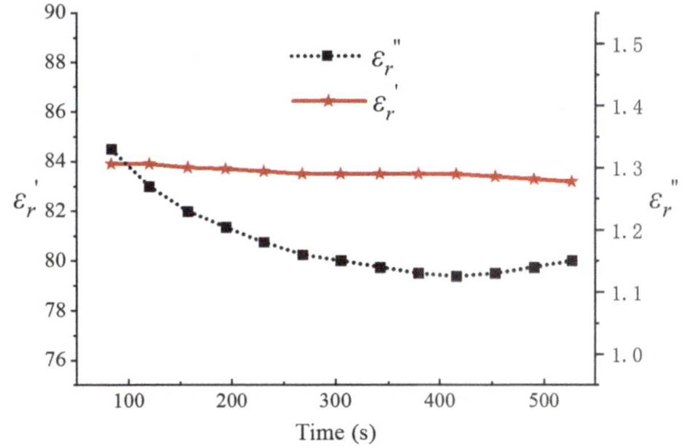

Fig. 2.9 Relative permittivity versus time of the reacted solution at 25 °C

A typical saponification reaction between $CH_5COOC_2H_5$ and NaOH in a dilute solution is selected for verifying this method's feasibility in a reaction system. The reaction equation is.

$$CH_5COOC_2H_5 + NaOH \rightarrow CH_5COONa + C_2H_5OH \tag{2.39}$$

From the measured reflection coefficient, the unknown relative permittivity of the dilute solution in the saponification reaction can be calculated by employing the GA-based inverse-calculation technique described in the previous section. The results of the calculated relative permittivity are shown in Fig. 2.9, which indicates that during the saponification reaction, the real part of the relative permittivity of the dilute solution remains almost constant, but its imaginary part changes significantly due to a significant change in the conductivity of the solution. In fact, the significant change in conductivity can be used to determine the reaction rate. The experimental/computational results show that the coaxial sensor is very sensitive to a change in the relative permittivity with time near the resonant frequency.

2.3.2.2 Neural Network Algorithms

Introduction

This section is concerned with the application of an advanced BP (Back Propagation) neural network optimization technique to measure the permittivity of materials and reaction systems.

Neural networks can be trained to learn the behavior of the relative permittivity of a material under microwave irradiation. When the network is sufficiently trained, it is supplied with the values of the measured complex S-parameters and determines ε_r' and ε_r'' of the sample. Unlike the GA that solves the optimal value, the neural network is mainly used to deal with the relation between the input and output. For a dielectric measurement, a GA is used to solve the permittivity optimal solution and a BP neural network is used to inverse the relative permittivity from the scattering parameter, avoiding the modeling and equation problems of other methods. In addition, a BP neural network can optimize the probe structure. Fast, accurate, and reliable neural network models can be developed from the measured/simulated microwave data.

Here, we are concerned with the application of an advanced BP neural network optimization technique to measure the permittivity of materials.

Theory and Application of Probe Design

Figure 2.10 shows a process for reconstructing the permittivity of a material by using a BP network. The result can be obtained very quickly once the network has been trained, because the samples have been produced by the FDTD method before being trained. Neural model development using simulation data has some advantages: first, it is relatively easy to sweep any parameter in the simulator, and second, any

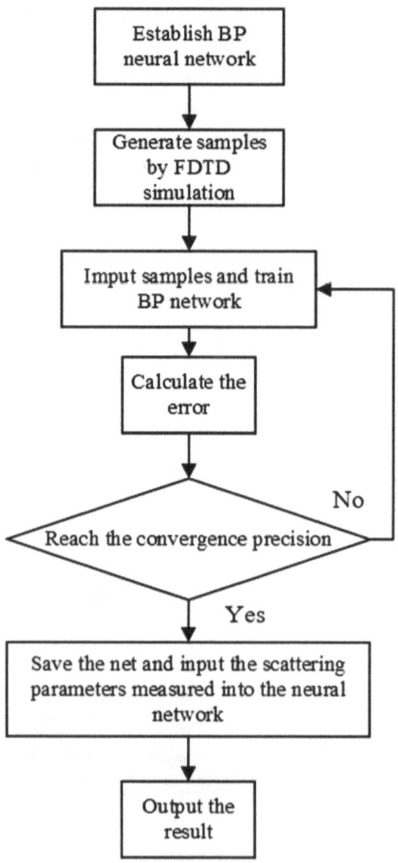

Fig. 2.10 Flowchart of the reconstruction of permittivity by BP

response can be modeled as long as it can be computed in the simulator. Training data are generated using a special procedure, which repeatedly operates the numerical simulation computing the reflection coefficient for different values of the permittivity of the sample.

The network is trained by samples before making the measurement, and then, the test data are obtained from the measurement. A training technique, namely back propagation, is implemented using the gradient method. When the network is sufficiently trained, it is supplied with the values of the measured S-parameters, and then, determines ε_r' and ε_r'' of the sample in several seconds. The test data can be processed in parallel, and many reconstructed results can be obtained simultaneously. Therefore, the online measurement is realized using a BP neural network.

Measurement of a Relative Permittivity of Solid

The measurement probe and simulation model are shown in Fig. 2.11. The coaxial probe is designed with a pointed-end structure for plugging into powders easily. It can measure the magnitude and phase of the reflection coefficients (|S11|, ψ_{S11}) at the frequency of interest, f_0, and in this instance, is contained in an iron can structure. The measurement apparatus can be used over a broad frequency band and in an increasing temperature environment.

In the process, data that are needed to train the networks, and their performance, are evaluated obtained from simulations accomplished by FDTD. When facing the original data from the simulation, the ranges of different inputs, such as the unit and magnitude, should be normalized. A BP neural network with normalization is measurably superior to the standard online BP algorithm in terms of convergence speed and accuracy. The input data normalization with certain criteria prior to a training process is crucial for obtaining good results as well as for significantly accelerating the calculations. The neural network is convergent using the normalization of the input data when the number of epochs attains 6000. The normalization affects the convergence properties of the neural network and exhibits a superior performance over the traditional method in terms of convergence speed. Then, anti-normalization must be used when the output data are obtained from the trained neural network.

Before constituting the model, the primary question is the production of samples. On one hand, the characteristics of the network should be considered; on the other

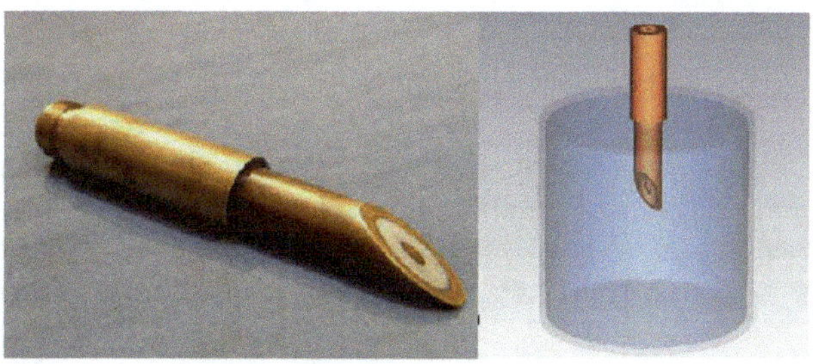

Fig. 2.11 Structure of the open-ended coaxial probe and the simulation model

Table 2.3 Effective permittivities of a range of materials at 2450 MHz

	Real part of relative permittivity			Imaginary part of relative permittivity		
Material	Measurement	Reference	Relative errors (%)	Measurement	Reference	Relative errors (%)
DMSO	46.890	48.900	−4.1	19.223	20.002	−3.9
Methanol	25.910	24.970	3.8	14.010	14.523	−3.5
Formic acid	6.839	7.135	−4.1	1.325	1.359	−2.5
Ethanol	8.680	8.939	−2.9	7.305	7.585	−3.7
Acetone	20.021	19.325	3.6	1.132	1.099	3.0
Glycol	19.576	18.969	3.2	17.489	16.898	3.5
Glycerin	6.909	7.123	−3.0	3.685	3.838	−4.0

hand, the sample data must closely reflect the intrinsic rule of the material's permittivities.

The learning strategy is implemented through the selection of the most informative training samples. The informativeness is defined as the sensitivity of the neural network output to perturbations in the input value of that pattern. The output sensitivity vector is defined as

$$S_o^{(p)} = \left\| S_{oz}^{(p)} \right\|_2 \tag{2.40}$$

where $S_o^{(p)}$ is the output sensitivity vector and $S_{oz}^{(p)}$ is the output−input layer sensitivity matrix. Each element $S_{oz\ ,ki}^{(p)}$ of the sensitivity matrix is defined as a single output unit, $o_{k,}$ to changes in the input vector, and k is the total number of output units.

$$S_{oz\ ,ki}^{(p)} = \frac{\partial o_k}{\partial z_i^{(p)}} \tag{2.41}$$

Now, the permittivity of the material can be obtained when the scattering parameters measured are substituted into the trained BP network. The trained network can use the scattering parameter to measure the relative permittivity of materials quickly. Table 2.3 shows the reconstructed results of several organic solvents at 2.45 GHz. Accurate reconstructed results are achieved using the BP network. Table 2.4 shows the reconstructed results of a saline solution at 1 GHz, 15 °C. The results show that the measurement apparatus can be used in a broadband measurement.

The reconstructed results of the effective permittivities of solvents through the ANN agree well with the previously published data. Compared to the measurement methods mentioned above, the measurement technique in this paper has some advantages. The measurement system can operate in a broad frequency band and

Table 2.4 Effective permittivities of NaCl solutions of different fractions at 1 GHz

ω (NaCl) (‰)	Real part of relative permittivity			Imaginary part of relative permittivity		
	Measurement	Reference	Relative errors (%)	Measurement	Reference	Relative errors (%)
5	82.72	79.85	3.5	16.87	17.57	−4.0
10	80.49	78.15	2.9	28.53	29.70	−3.9
20	77.75	74.89	3.7	54.70	53.17	2.9

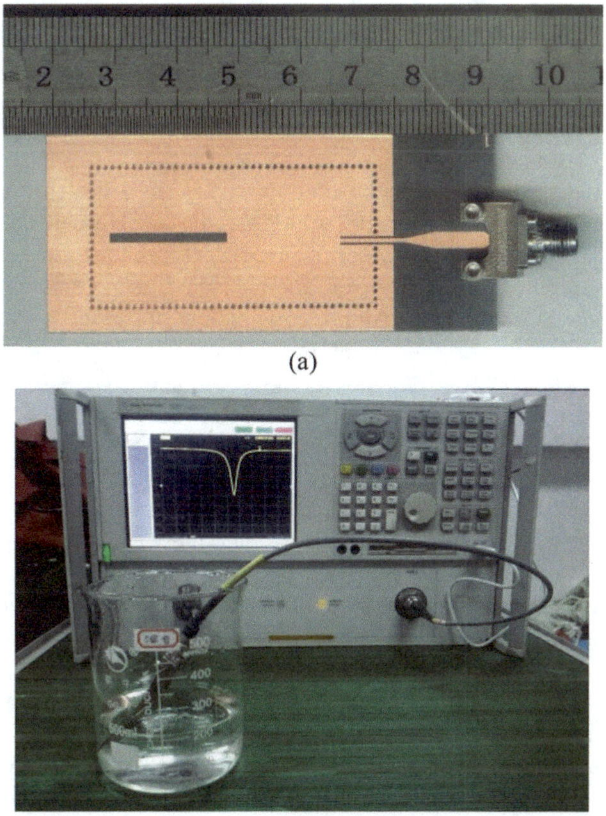

(a)

(b)

Fig. 2.12 The SIW resonator sensor and the measurement system. (**a**) The fabricated SIW resonator sensor. (**b**) The measurement system

deal with rising temperatures; the sample may be a liquid, powder, or gas. Moreover, the neural model developed using simulation data is very convenient.

Measurement of Liquid Relative Permittivity

To measure the relative permittivity of liquids, we present a novel Substrate Integrated Waveguide (SIW) resonator sensor (Fig. 2.12) that is specifically designed from experimental data and ANN. The resonant characteristics of the sensor are influenced by liquids through an open slot on the top plane. The inverse problem of obtaining permittivity is solved using an ANN. Experiments are performed with an SIW resonator sensor designed at the C band. Experimental data are reported and agree well with the reference values. The sensor is simple and low-cost, which may be applied to permittivity measurements in applications at the industrial, scientific, and medical frequencies.

For measurement over a broad permittivity range, the binary liquid mixture of ethanol and water is chosen. The SIW resonator sensor is immersed in the liquid under test in a 500 mL glass beaker at 20 °C. The S_{11} parameters of the SIW resonator sensor immersed in the binary liquid mixtures of different volume fractions are measured from 5.75 to 5.95 GHz by using an Agilent N5230A VNA. During the measurement, the average function of VNA is used to reduce random errors. Then, the resonant frequency and unloaded quality factor are extracted from the measured S-parameters.

Considering that the random errors from measuring the volume of liquid and variations in temperature will affect the relative permittivity, we analyze the uncertainty of the relative permittivity measurement. The extracted resonant frequency and unloaded quality from the measured parameters under different conditions are shown in Fig. 2.13. An effective method for reducing the random errors originating from measuring the liquid is achieved by multiple measuring (3–4 times) to obtain average values. Maintaining a constant temperature is also important for an accurate measurement.

Figure 2.14 shows the real and imaginary parts of the reconstructed relative permittivity. They agree with the values calculated according to the dielectric relaxation parameters of ethanol–water mixtures with different volume fractions. The average relative errors of measured and calculated results are 5% and 7%, respectively. The proposed SIW resonator sensor is sensitive to the relative permittivity variation of the liquid under test. It can figure out the valley of the unloaded quality factor curve when the volume fraction of ethanol is between 0.8 and 1. However, the inaccuracy of the simulated data will cause higher relative errors when reconstructing the relative permittivity, especially if the condition cannot be simulated accurately when ethanol begins to be added to pure water. An effective solution is to use the SIW resonator sensor to measure different types of liquids whose relative permittivities are known, and then use these measured data to train an ANN instead of the simulated data. Thus, more accurate reconstructed results will be obtained due to the exclusion of the simulated errors.

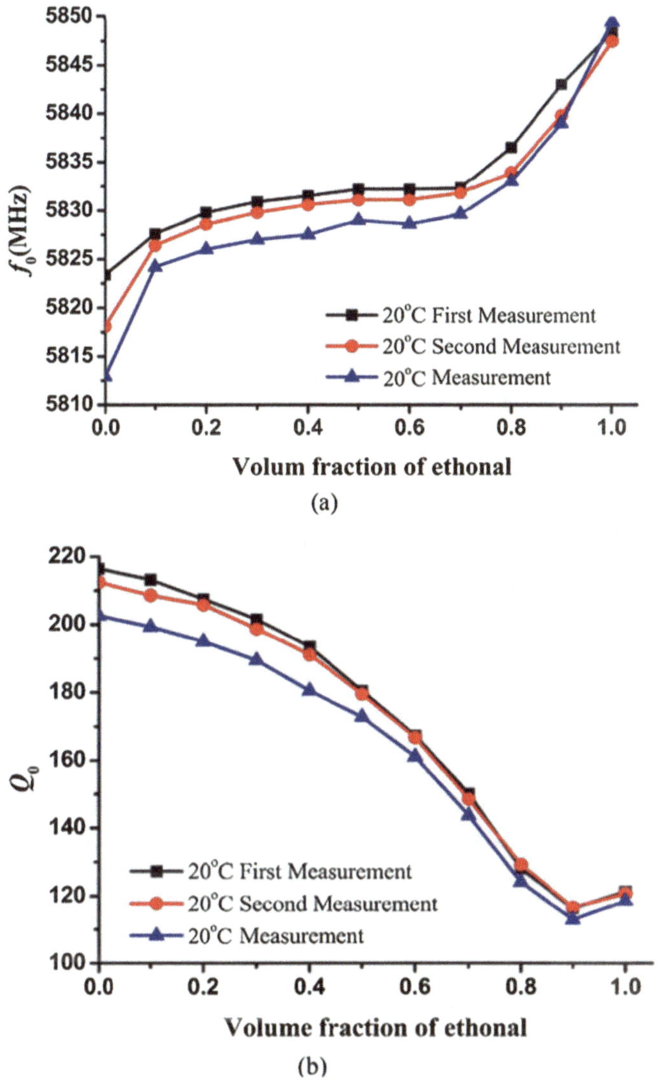

Fig. 2.13 Measured results with respect to volume fraction of ethanol from 0% to 100%. (**a**) Resonant frequencies. (**b**) Quality factors

2.4 Characterization and Measurement of Relative Permittivity of Chemical Reactions

Generally, the relative permittivity of reactants changes with not only temperature but also reaction time. In the simulation of microwave heating of chemical reactions, we must know the relative permittivity as a function of temperature and reaction

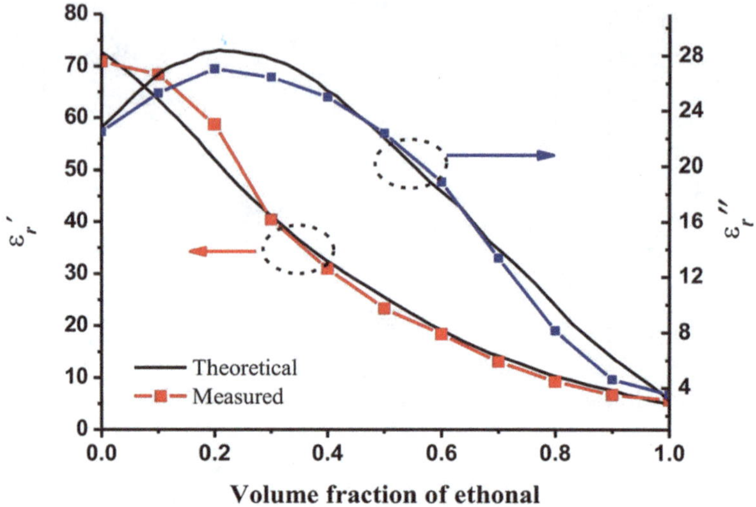

Fig. 2.14 Reconstructed relative permittivity (real and imaginary parts) with respect to the volume fraction of ethanol from 0% to 100%

time. Huang presented a computational method for the calculation of the effective complex permittivity of a heterogeneous material using its microscopic properties and structure [2]. This new method for calculating the effective properties aims to guarantee that the electric energies stored and dissipated by the heterogeneous material are the same. The analysis of a bi-phase mixture with a periodic cubic sample is tested to check the homogenization of the material. An iterative procedure that corrects the initial permittivity yields an electric field distribution of the effective material, which is in good agreement with the electric field of the real mixture.

2.4.1 Huang Method for the Iodination of Acetone

The acetone iodation reaction is widely used to study the chemical kinetics since significant change in relative permittivity can be observed. Here, the acetone iodation reaction is carried out at 2450 MHz to verify the reaction equation, as follows:

$$KIO_3 + 5KI + 6HCl + 3CH_3COCH_3 ® 3CH_3COCH_2I + 6KCl + 3HI + 3H_2O. \quad (2.42)$$

The initial concentrations are $C^0_{KIO_3} = 0.005$ mol/L, $C^0_{KI} = 0.025$ mol/L, $C^0_{HCl} = 1.5$ mol/L, and $C^0_{CH_3COCH_3} = 1.5$ mol/L. The reaction is carried out at four constant temperatures (30 °C, 37 °C, 44 °C, and 52 °C). If we assume that the

Table 2.5 Coefficients of the constants for the acetone iodation reaction: the coefficients of the constants to calculate the real part of relative permittivity

i	Real part relative permittivity		
	Ai	Bi	Ci
0	13.02656	−0.02987428	57.18999
1	984.429928	−2.91099666	−9.92749
2	57566.693328	−199.02838	−0.02172
3	−3312088.858	10959.69268	
4	61429666.84	−20169.96 8	
5	−589243238.7	1929467.137	
6	31,776,248 52	−10395366.6	
7	−9,168,741,336	29983096.4	
8	11,030,425,780	−36063797.5	

Table 2.6 Coefficients of the constants for the acetone iodation reaction: the coefficients of the constants to calculate the imaginary part of relative permittivity

i	Imaginary part relative permittivity		
	Ai	Bi	Ci
0	−1.75210	0.010061	10.0413
1	1833.84832	−6.44949	6.74988
2	−218986.135	748.27307	−0.02508
3	6710456.844	−22739.5615	
4	−104509777.5	352572.4053	
5	938580726.3	−3157607.49	
6	−4,912,201,929	16495354.14	
7	1,391,981,230	−46683237.4	
8	−16,507,662,920	55310783.14	

concentration of the consumed acetone is x after t seconds, the rate equation can be written as

$$-\frac{d\left(C^0_{CH_3COCH_3}-x\right)}{dt}=K_{rate}\cdot\left(C^0_{CH_3COCH_3}-x\right)\cdot\left(C^0_{H^+}+x\right). \tag{2.43}$$

By integrating Eq. (2.43), we have

$$x=\frac{C^0_{CH_3COCH_3}\cdot C^0_{H^+}\cdot\left(e^{\left(C^0_{CH_3COCH_3}+C^0_{H^+}\right)K_{rate}\cdot t}-1\right)}{C^0_{CH_3COCH_3}+C^0_{H^+}\cdot e^{\left(C^0_{CH_3COCH_3}+C^0_{H^+}\right)K_{rate}\cdot t}} \tag{2.44}$$

Therefore, the concentration of the consumed acetone can be obtained. According to the obtained results, the three measured curves at 30 °C, 44 °C, and 52 °C are selected to calculate the coefficients of the constants. The results are presented in Tables 2.5 and 2.6.

After these coefficients are obtained, the relative permittivity of the acetone iodation reaction at any concentration and temperature can be deduced from Fig. 2.15.

The measured and calculated results for temperature and concentration are shown in Figs. 2.16 and 2.17, respectively.

For multiphysics calculations, finding how the relative permittivity of the reaction mixture changes with the reactant concentration and temperature is preliminary and necessary. Here, we improve the method to obtain the formula, which can be directly applied to the multiphysics calculation for the microwave heating of chemical reactions. This method enables to analyze the nonlinear time-dependent changes in reflection and absorption of the microwaves by the reactants and can instruct the design of microwave chemical reactors.

Most chemical reactions are sensitive to temperature; therefore, using microwaves to heat reactants presents an impressive potential application. Generally, the relative permittivity of reactants changes with not only temperature but also reaction time. In the simulation of the microwave heating of chemical reactions, we need to know the relative permittivity as a function of temperature and reaction time. Huang proposed a method based on the experimental results to calculate the relative permittivity of chemical reaction systems as a function of temperature and reaction time.

2.4.2 *Gaussian Function for Biodiesel Reaction*

2.4.2.1 Gaussian Equation

For a liquid chemical reaction, the relative permittivity of the reacting mixture solution can be determined by the components and the temperature of the reaction solution. The components of the reaction solution can be described by the concentration of one reactant; consequently, the relative permittivity can be described by a bivariate function of the temperature and concentration of one reactant. On the other hand, a nonlinear model, for example, the Gaussian function, has the potential to better fit nonlinear data with a few coefficients. Therefore, the relative permittivity of a reaction mixture solution is expanded to the power series of the concentration and exponential function of temperature. The formula can be expressed as follows:

$$\varepsilon'_{eff} = \varepsilon_0 \varepsilon'_r = \varepsilon_0 \left\{ \sum_{i=0}^{N} a_i C^i + d_0 \exp\left[-\left(\frac{T - d_1}{2d_2} \right)^2 \right] \right\} \tag{2.45}$$

$$\varepsilon''_{eff} = \varepsilon_0 \varepsilon''_r = \varepsilon_0 \left\{ \sum_{i=0}^{N} b_i C^i + g_0 \exp\left[-\left(\frac{T - g_1}{2g_2} \right)^2 \right] \right\} \tag{2.46}$$

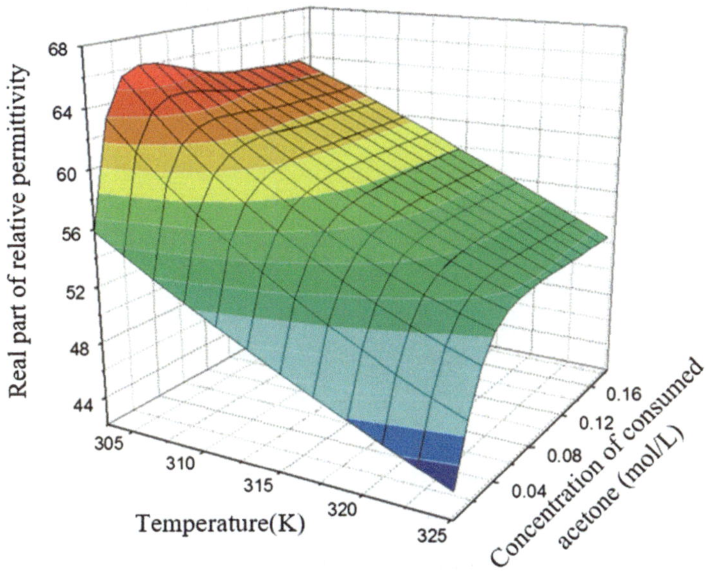

(a)

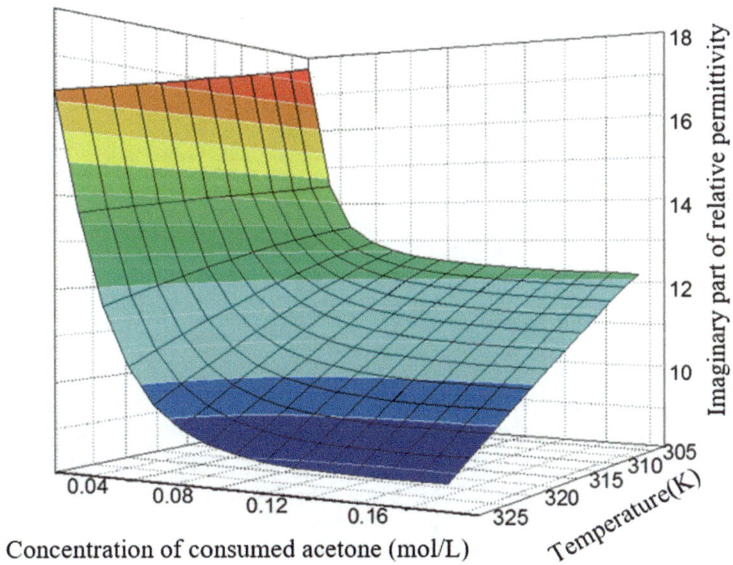

(b)

Fig. 2.15 Relative permittivity of the acetone iodation reaction against concentration and temperature: (**a**) Real part of relative permittivity, (**b**) imaginary part of relative permittivity

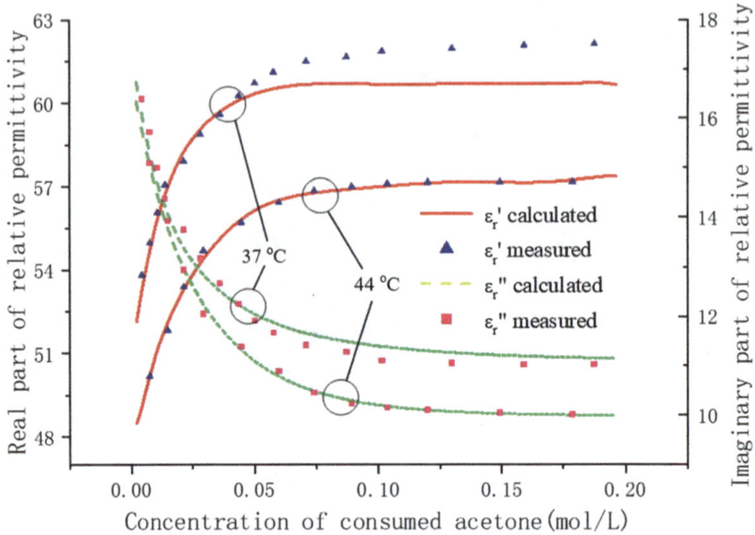

Fig. 2.16 Comparison of the calculated and measured results against concentration

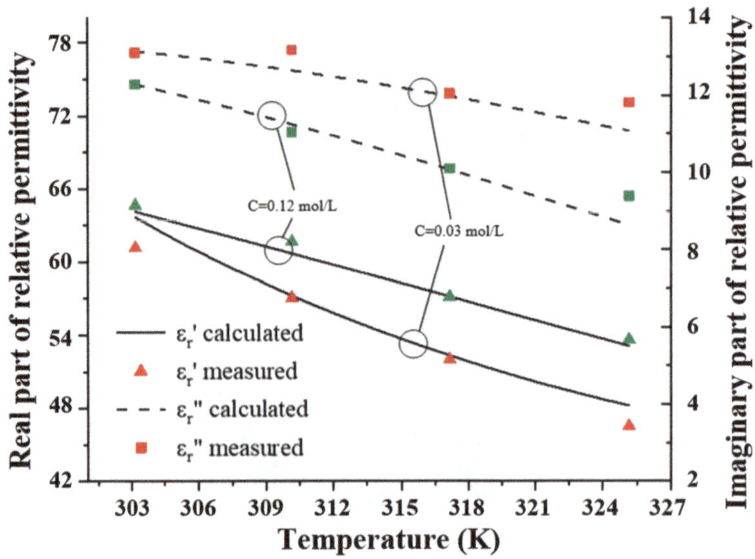

Fig. 2.17 Comparison of the calculated and measured results against temperature

where ε'_{eff} and ε''_{eff} are the real and imaginary parts of relative permittivity, respectively. T indicates the temperature of the reaction mixture solution. C indicates the concentration of one reactant in the reaction solution. a_i, b_i, d_i, and g_i are unknown coefficients. Usually, $N = 2$ is enough to fit the change in the relative permittivity with the concentration of one reactant. In this case, six coefficients need to be determined in Eq. (2.45). a_0, a_1, a_2, d_0, d_1, and d_2 are used to calculate the real part of relative permittivity, while b_0, b_1, b_2, g_0, g_1, and g_2 are used to calculate the imaginary part. Here, the measurement is innovatively implemented in a temperature cycle manner and the measurement of relative permittivity over a wide temperature range. The curve of relative permittivity with time and temperature can be obtained via one measurement. By using the reaction rate equation, the measured result can be transferred into a new curve against the concentration of one reactant and temperature. Then, the data in Eq. (2.45) are substituted and used to calculate the coefficients a_i, b_i, d_i, and g_i via least-square fitting by MATLAB.

2.4.2.2 Biodiesel Reaction

Biodiesel is produced by the esterification of fatty acids and methanol. As a typical fatty acid, oleic acid is employed to produce biodiesel with methanol. The catalyst in the reaction is concentrated sulfuric acid. The following is the reaction equation:

$$RCOOH + CH_3OH \underset{k_{-1}}{\overset{k_1}{\rightleftarrows}} RCOOCH_3 + H_2O \tag{2.47}$$

As for the synthesis of biodiesel, the molar ratio of methanol to oleic acid is 6:1, and the mass fraction of concentrated sulfuric acid to the whole reaction solution is 1%. The reaction rate equation can be written as follows:

$$-\frac{d\left(C^0_{RCOOH} - C_{H_2O}\right)}{dt} = k_1\left(C^0_{RCOOH} - C_{H_2O}\right)\left(C^0_{CH_3OH} - C_{H_2O}\right) - k_{-1}C^2_{H_2O} \tag{2.48}$$

where C^0_{RCOOH} and $C^0_{CH_3OH}$ are the initial concentrations of oleic acid and methanol, respectively; C_{H_2O} indicates the concentration of water; k_1 is the rate constant for the forward reaction; and k_{-1} is the rate constant for the reverse reaction; the rate constants satisfy the Arrhenius equations:

$$k_1 = A_1 \exp\left(-\frac{E_a^+}{RT}\right) \tag{2.49}$$

$$k_{-1} = A_{-1} \exp\left(-\frac{E_a^-}{RT}\right) \tag{2.50}$$

where E_a^+ and E_a^- are the activation energies of the forward and reverse reactions, with values 33.8 and 13.9 kJ/mol, respectively. The pre-exponential factors A_1 and A_{-1} are 743.6 and 0.606, respectively. R is the universal gas constant.

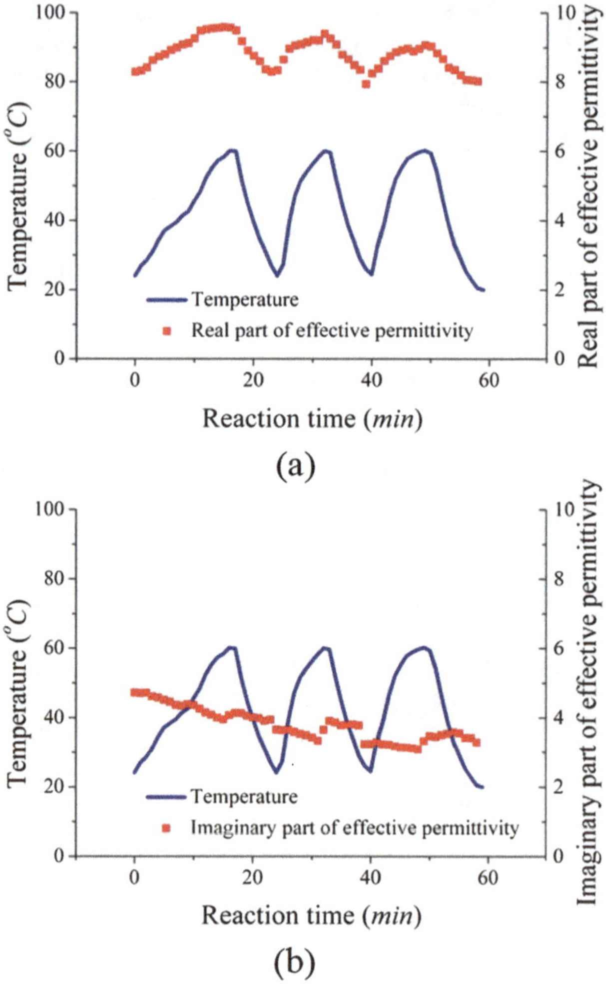

Fig. 2.18 Measured temperature and relative relative permittivity of the biodiesel reaction. (**a**) Real part of relative permittivity. (**b**) Imaginary part of relative permittivity

Therefore, the concentration of water in the reaction can be calculated. Then, the measurement results as a function of time can be transferred as a function of the product's concentration.

The boiling point of methanol is 64.8 °C under standard atmospheric pressure, so the bubbles will affect the scattering parameter of the solution. Therefore, the reaction is conducted over a temperature range from 20 °C to 60 °C. The temperature and measured relative permittivity with time are shown in Fig. 2.18. The relative permittivity at different product concentrations and temperatures can be obtained by Eq. (2.49). Then, by combining with Eq. (2.48), the coefficients in the equation can be obtained via least-square fitting. Finally, the relative permittivity at any temperature and concentration can be obtained.

2.4.2.3 Experimental System and Method

An Agilent E8362B microwave vector network analyzer is used to measure the reflection coefficients from a specially designed coaxial probe in the measurement system, which is immersed into the reaction solution. The probe's location must be the same as in the mathematic model when we calculate the ANN samples by using the FDTD method. The measurement is performed at 2450 MHz, which is the globally used ISM band. A well-designed 500-ml three-necked florence flask is utilized. At the same time, a UMI-8 optical fiber thermometer is employed to measure the temperature of the solution. The 85–2 magnetic stirrer's rotation speed in the experiment is 2000 r/min. A customized water bath trough is used to control the temperature of the reaction solution in the flask by exquisitely controlling the temperature and speed of the influent water. In this way, a temperature curve can be obtained as expected; meanwhile, the scattering coefficients of the reaction system are measured online by using the coaxial probe.

The ANN computational module has been widely used in microwave technology recently. Here, the ANN model based on a BP training algorithm is employed to inverse the relative permittivity. First, a measurement system is designed and the scattering parameter at 2450 MHz is calculated by using the frequency-dependent FDTD method. Second, a BP nerve network and enough simulated samples are combined to train the network. Then, the trained BP neural network is employed to reconstruct the relative permittivity by using the measured scattering parameter, and the inverse process is shown in Fig. 2.19.

2.4.2.4 Results

The relative permittivity of the reaction solution is a bivariate function of the reaction's components and temperature. Here, we describe it as a 3D plot, which clearly shows the change trend of the concentration of one reactant and temperature. Figure 2.20 shows the comparison between the calculated and measured effective permittivities for the synthesis of biodiesel against concentration and temperature. In the figures, the red points represent the measured results, which are larger than the

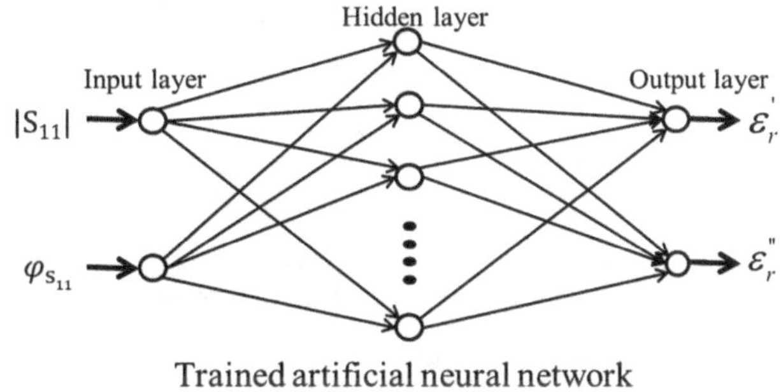

Fig. 2.19 Inversion of relative permittivity by a trained network

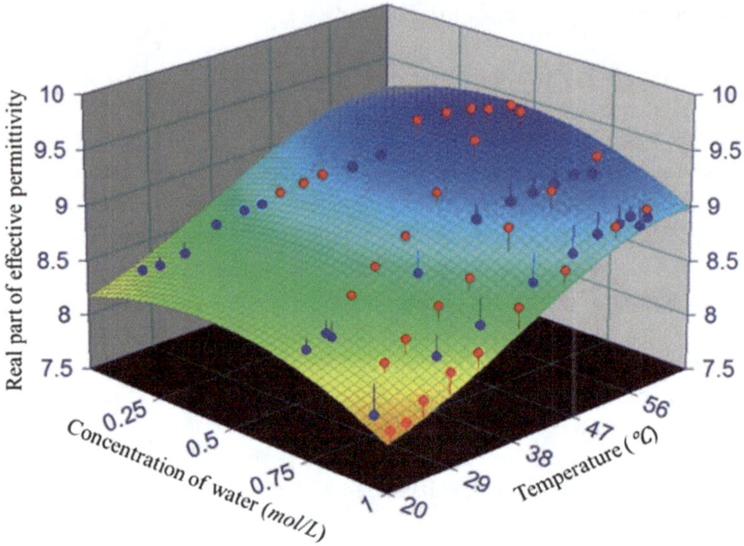

(a)

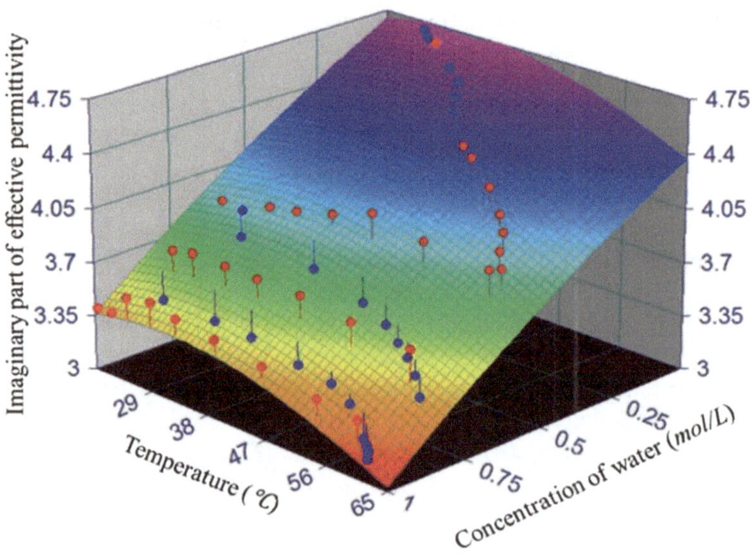

(b)

Fig. 2.20 Comparison between the calculated and measured effective permittivities of the synthesis of biodiesel with concentration and temperature at 2450 MHz. (**a**) Real part of relative permittivity. (**b**) Imaginary part of relative permittivity. (Note: The points are the measured results, and the surface is the calculated result)

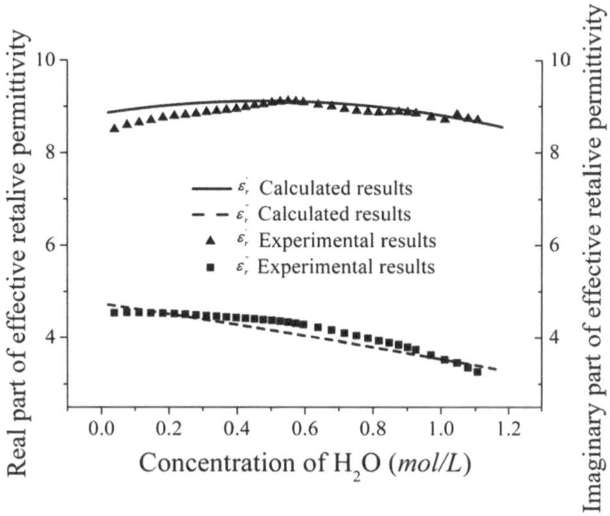

Fig. 2.21 relative permittivity changes with water concentration at a constant temperature of 44 °C

corresponding calculated results, while the blue points represent the smaller ones. As expected, the figures reveal that the real and imaginary parts of relative permittivity change significantly with concentration and temperature. The nonlinear change in the real part of relative permittivity with time can be found. The real part of relative permittivity increases with time after the reaction starts, but then begins to decrease after a short period and the reaction solution becomes muddy. The imaginary part of relative permittivity basically decreases during the whole reaction period.

As for the calculation of the real part of the synthesis of biodiesel, relative errors of the calculation in most of the measuring points are less than 1.5%, with a maximum relative error of 4.10%; while for the imaginary part, errors in most of the measuring points are less than 3%, with a maximum of 5.31%. Meanwhile, the biodiesel reaction under a constant temperature of 44 °C is carried out to validate the calculated results. The calculated and measured effective permittivities are compared with a water concentration at a constant temperature of 44 °C in Fig. 2.21. The maximum relative error of the real part is 4.10% and that of the imaginary part is 5.65%.

The comparisons between the calculated and measured results shown in Figs. 2.20 and 2.21 indicate that the accuracy of the surface fitting obtained through Eq. (2.48) is pretty good. However, for the imaginary part of the relative permittivity, the difference between the calculated and measured results is relatively large compared to the real part. This phenomenon is largely because the values of the imaginary part of relative permittivity are smaller than those of the real part. Moreover, the imaginary part of relative permittivity is usually more sensitive to temperature. Therefore, the difference between the calculated and measured values of the imaginary part of relative permittivity is relatively larger than that of the real part. Clearly, the calculated results are in good agreement with the measured results.

2.5 Summary

In summary, we introduced the concept and properties of permittivity of materials. Then we propose dielectric properties and some special phenomena between microwave and chemical reaction systems, which are not consistent with the traditional microwave theory. For these phenomena, we propose some methods for measuring relative permittivity in chemical reaction system. Finally, taking biodiesel reaction as an example, we introduce the dielectric properties of the reaction from theory, design and experiment.

References

1. Holland JH. Adaptation in natural and artificial system. Ann Arbor: The University of Michigan Press; 1975.
2. Huang K, Yang XQ. New method to measure and calculate the rate constant of an acetone iodation reaction. Ind Eng Chem Res. 2005;44:4501–3.

Chapter 3
Dynamic Analysis of Microwave Heating of Chemical Reactions

3.1 Challenges to Microwave Energy Application

3.1.1 *Nonuniformity Heating*

Microwave heating is fundamentally nonuniform. As the energy penetrates a lossy material, part of it is absorbed and the other penetrates further, and so energy absorption is nonuniform. The properties of the load, size, and geometry, as well as the design of the microwave oven, complicate this scenario of energy absorption, but nonuniformity remains the rule. The composition and temperature variation in a chemical reaction will change during the heating process, which causes the dielectric properties of the load to vary during heating and contributes to the spatially nonuniform energy absorption.

In general, the in-homogeneity of microwave heating is harmful in applications, although sometimes we can take advantage of this characteristic of microwave heating, for example, different components of the reaction products can be obtained by the nonuniformity of microwave heating. Numerous methods have been suggested to improve heating uniformity in microwave ovens, including the use of microwaves in combination with other heating sources, changing the load composition, and using active packages such as shields and susceptors. In addition, some ovens include mode stirrers to continuously change the electromagnetic field to obtain uniform heating or a turntable to change the position of the heated object during the heating process to obtain uniform energy absorption. However, no specific criteria have evolved for increasing the temperature uniformity of a load being heated in a microwave oven. This is partly due to the complex nonlinear physics governing microwave heating, but more significantly because various loads (with different properties, shapes, and volumes) need different specifications.

K. Huang et al., *Dynamics in Microwave Chemistry*,
https://doi.org/10.1007/978-981-15-9655-1_3

3.1.2 Thermal Runaway

Thermal runaway is a type of thermal instability due to the interaction between electromagnetic waves and materials. In other words, it is the result of the positive feedback caused by the interaction between microwaves and reaction systems, as illustrated in Fig. 3.1. A slight change in the input power may cause a significant variation in temperature in the reaction system due to thermal runaway. When thermal runaway occurs, the reactants may be burned out and the microwave reactor may be damaged; even an explosion may occur. This seriously hampers the industrial application of microwaves for heating in the chemical industry.

The uniformity of microwave heating and the efficient use of energy require optimization analysis through multiphysics coupling calculations to achieve the expected application results.

3.1.3 Scale-Up

The process of heating involves almost all sectors of the national economy, which is widely used in national productivity and people's daily lives. As a new technology, microwave heating has many advantages over other heating methods. It will, therefore, undoubtedly be promoted and applied in various areas. However, microwave heating also has some disadvantages. For example, when microwave heating technology is used to heat a material of large volume, if the conductivity and dielectric constant of the heated material are large, the ability of microwave to penetrate the material is poor, and the penetration depth is given as

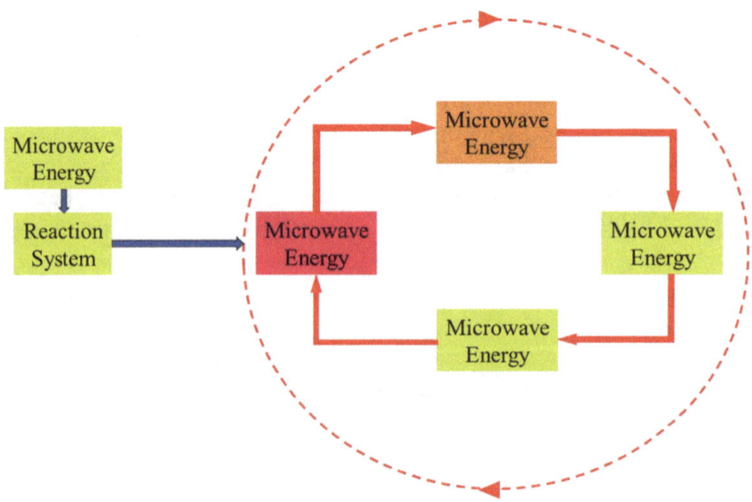

Fig. 3.1 Positive feedback of microwave heating

$$\delta = \sqrt{\frac{2}{\omega\mu\sigma}} \tag{3.1}$$

where δ is the penetration depth, ω is the microwave angular frequency, μ is the relative permeability, σ is the electrical conductivity. The result of this situation is that the outer surface of the material is heated and the interior is not successfully heated. The energy used in microwave drying is high-priced electric energy, which compared to traditional energy, is still high. The microwave drying of materials alone tends to cause an excessive temperature increase in the material, and the high temperature, if not properly controlled, may result in "burst" in the material, and even scorching.

In the 1990s, due to the rapid development of electronic technology, microwave heating technology became increasingly mature and the equipment became better. In addition, the power supply was greatly improved and drastically reduced the direct cost of microwave heating equipment. The deteriorating global environment has made people realize that traditional heating is not environmentally friendly. This provides a good opportunity and broad prospects for the application and development of microwave heating. We foresee that microwave heating technology, with its unique advantages, will play a very important role in future production. In conclusion we need to use technology to overcome these problems.

3.2 Multiphysics Computation

3.2.1 Introduction to Multiphysics Computation

In nature, there are four types of physical fields: displacement, electromagnetic, temperature, and fluid fields. To be able to describe some phenomena induced by these fields, we must apply some basic theory of physics, chemistry, and mathematics to build the right model. The physical process and its results can then be acquired. In most cases, two or more physical fields should be coupled to describe the same physical process, because these physical fields can affect each other. This computation is usually called multiphysics computation. In terms of multiphysics simulation, some parameters of the physical fields in the partial differential equations can be coupled.

Many commercial software packages can integrate many physical, chemical, and mathematical modules. In this book, a software which based on FEM is introduced. And their products were expanded to include some special modules, including structural mechanics, high- and low-frequency electromagnetic fields, fluid flow, heat transfer, chemical reactions, MEMS, and acoustics. Based on these modules, the software released some products that synchronously link with other software: CAD, MATLAB, and Excel, which has created a platform for multiphysics computation.

Many numerical calculation methods can be applied to multiphysics computations, such as FDTD, finite-element method (FEM), finite-volume method (FVM), and the boundary element method (BEM). In this book, which software used is based on FEM for the computations. Here, we give a simple introduction of FEM, which is a type of numerical technology to approximately work out mathematical boundary value problems. It was proposed in the 1940s and was applied to the design of aircraft. The basic principle of this method is to replace the continuous domain with many sub-domains. In these sub-domains, unknown functions can be represented by simple functions with unknown coefficients. Therefore, the original boundary problem with infinite degrees of freedom is changed to that with finite degrees of freedom. In other words, the solution of the whole system is approximated by finite unknown coefficients. The Rayleigh-Ritz method* or Galerkin* method is then used to obtain a set of algebraic equations. Finally, the solution to the boundary value problems can be obtained by solving these algebraic equations. The Rayleigh-Ritz method* and Galerkin* method will not be described here in detail; however, readers can consult the book "FEM in Electromagnetics" by Jianming Jin [1].

3.2.2 Challenges to the Multiphysics Computation of Microwave Heating for Industrial Application

Multiphysics computation is a good method to design microwave chemistry reactors; however, large challenges exist in the computation.

1. The dielectric properties of a non-balanced system

   In general, most materials or substances are in equilibrium state. Nevertheless, the chemical reactions are non-equilibrium state systems, and thus, we cannot describe their dielectric properties using permittivity or the dielectric function. In the following sections, two vectors are derived to describe the dielectric properties.
2. Meshing problem

   Mesh generation is always a key step of multiphysics computations. Particularly, when moving elements exist in the model, we should dynamically consider the meshes. Several mesh technologies will be introduced below.
3. Coupling of different physical parameters

   In a microwave chemical system, the interaction between microwaves and chemical reactions is an extremely complex process, which will inevitably result in the calculation of spatial and temporal coupling at different scales. To simulate this process accurately, we propose an extended multiscale time-domain finite element method. The computational cost is huge when solving the coupling physical parameters in a multiphysics model. Therefore, a multigrid method, a time/frequency-mixing method, and other advanced calculation methods are adopted in the algorithm.

4. Large-scale calculation

   Industrial microwave applicators are usually large compared to the microwave wavelength, which brings a large cost for the electromagnetic calculation, including the mesh number and several degrees of freedom. Here, we give an example of calculation of the temperature rise on the Earth.

This chapter takes individual experiments as an example to explore the settings and solutions for multiphysics calculations from the aspects of dielectric properties, meshing, and the large-scale computation of non-equilibrium systems.

3.3 Algorithms for the Dynamic Analysis of Microwave Heating

When using microwaves for heating analysis, the algorithm needs to be studied first, and then, the algorithm of microwaves in the heating of solids and fluid will be introduced.

3.3.1 Multiphysics Simulation of Microwave Heating of Solid Materials

In the simulation software case library, the use of a microwave cavity to heat potato pieces was taken as an example to explain microwave heating of solid materials.

3.3.1.1 Geometry

In the multiphysics simulation software, the microwave oven model shown below was constructed and the heated solid material was a potato (Fig. 3.2).

3.3.1.2 Governing Equations

1. Electromagnetic field equation

   For the multiphysics calculation of the microwave heating of oleic acid, the electromagnetic field, the heat in the fluid, and the free and porous media flow modules are coupled with each other.

   The electromagnetic field is given by Maxwell's equations:

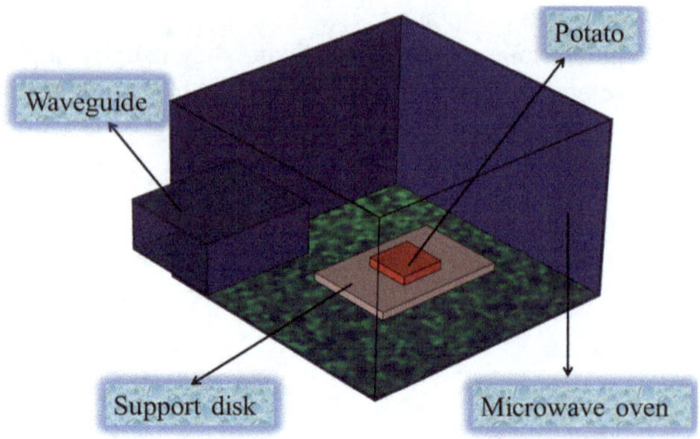

Fig. 3.2 Geometry of the microwave oven, potato, and waveguide feed

$$\begin{aligned} \nabla \times \vec{H} &= \vec{J} + \varepsilon \frac{\partial \vec{E}}{\partial t} \\ \nabla \times \vec{E} &= -\mu \frac{\partial \vec{H}}{\partial t} \\ \nabla \cdot \vec{B} &= 0 \\ \nabla \cdot \vec{D} &= \rho_e \end{aligned} \tag{3.2}$$

where $\vec{E}$ and $\vec{H}$ are the electric and magnetic field intensities, respectively, $\vec{D}$ is the electric displacement, $\vec{B}$ is the magnetic flux density, and $\vec{J}$ is the conduction current. ε is the relative permittivity of the substance, μ is the relative permeability, and ρ_e is the free charge density.

2. Power loss computation

   For general substances, the electromagnetic power loss, $Q_{\varepsilon,}$ can be obtained from the computed electric field by Eq. (3.3), where ε_0 is the permittivity of free space and ε'' is the imaginary part of the relative permittivity of the material.

$$Q_e = \frac{1}{2} \omega \varepsilon_0 \varepsilon'' \left| \vec{E} \right|^2 \tag{3.3}$$

3. Temperature distribution computation

   To compute the temperature distribution situation where only the potato slice in the model gets involved, the governing equation for heat transfer in solids is given as

$$\rho C_p \frac{\partial T}{\partial t} - k_c \nabla^2 T = Q = Q_e \tag{3.4}$$

where ρ is the material density, C_p is the material heat capacity under atmospheric pressure, T is the temperature, t is the time, Q is the heat source, and k_c is the thermal conductivity.

4. Boundary condition

(a) Electromagnetic boundary condition

The total surfaces of the heating system, namely the oven and the waveguide, are defined as perfect electric conductors, except for the microwave feeding port. The governing equation can be expressed as

$$n \times \vec{\mathrm{E}} = 0 \tag{3.5}$$

where n is the unit normal vector of the corresponding surface.

(b) Thermal boundary condition

- Thermal insulation

As the heat transfer part of the model only takes the processing materials into consideration, the thermal insulation boundary conditions, which indicate that there is no heat flux across the boundaries, are applied to define the whole surface of the heating material. The governing equation is defined as

$$-n \cdot Q = 0 \tag{3.6}$$

where Q is the heat flux.

- Thermal convection

As the heated object in the model will generate heat convection with the surrounding air. The thermal convection equation is defined as

$$Q_{\mathrm{loss}} = \Delta h \cdot (T - T_a) \tag{3.7}$$

where Δh is the thermal convection area, T_a is the ambient temperature.

5. Input parameters

Property	Applied domains	Value
Relative dielectric constant	Air	1
Relative permittivity	Potato	$\varepsilon(T)$
Relative permeability	Air Potato	1 1
Conductivity (S/m)	Air Potato	0 0
Heat conductivity coefficient (W/m K)	Potato	0.648

(continued)

Property	Applied domains	Value
Density (kg/m^3)	Potato	1050
Heat capacity at constant pressure (J/kg K)	Potato	3640

$$\varepsilon(T) = \varepsilon' - j\varepsilon'' = \frac{56.25 \cdot A + 16.57}{12T} + \frac{\sqrt{(56.25 \cdot A + 16.57)^2 + 2178 \cdot T^2}}{12T}$$
$$A = 1186.78 \cdot e^{\frac{2.88\times10^{-21}}{1.38065\times10^{-23}\cdot T}} \tag{3.8}$$

where ε' and ε'' are the real and imaginary parts of relative permittivity.

3.3.1.3 Results and Discussion

The calculated electric field distribution is shown in Fig. 3.3.

The surface temperature distribution of the potato is calculated as shown in Fig. 3.4.

Figure 3.3 shows that the electric field distribution in the heating chamber is relatively uniform. From the heated material in Fig. 3.4, after a different time of microwave heating, the warming effect is obvious and the highest surface temperature reaches 60 °C.

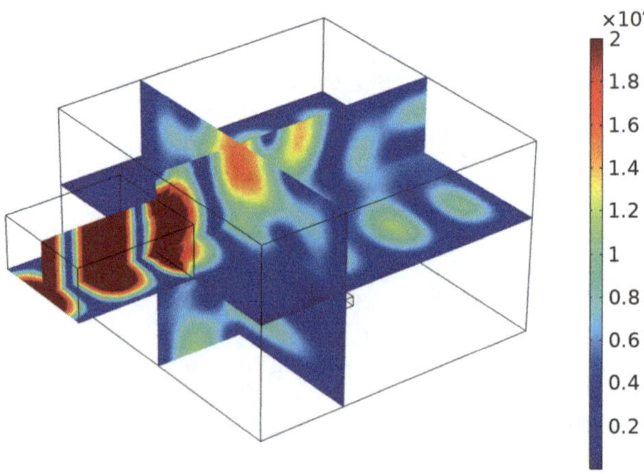

Fig. 3.3 Electric field distribution (V/m)

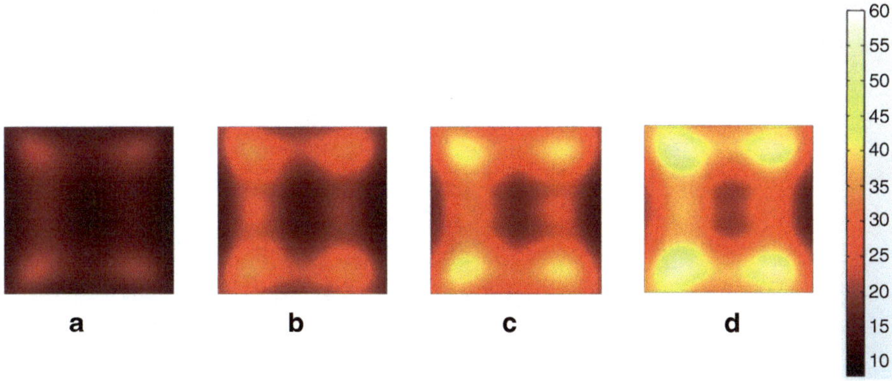

Fig. 3.4 Temperature distribution after (**a**) 5 s, (**b**) 10 s, (**c**) 15 s, and (**d**) 20 s of heating (C)

3.3.2 Multiphysics Simulation of the Microwave Heating of Fluid Materials

Microwaves are often used to heat fluids, and in this section, we take the microwave heating of water as an example to introduce the algorithm analysis of the microwave heating of a fluid. In addition, in this example, a single-mode cavity is used to calculate the fluid.

3.3.2.1 Geometry

The following simulation model was built using the multiphysics simulation software. A container is placed in the BJ22 waveguide, and the liquid placed in the container is water (Fig. 3.5).

3.3.2.2 Governing Equations

Currently described is fluid heating. The electromagnetic boundary conditions and thermal boundary conditions are consistent with the microwave heating potatoes in the previous section. For fluid modules, the flow in the free channel is described by the Navier–Stokes equations:

$$\begin{gathered}\nabla \cdot \left[\mu\left(\nabla\vec{u}\cdot\vec{I}\right)+\left(\nabla\vec{u}\right)^{T}-p\cdot\vec{I}\right]=\rho\left(\vec{u}\cdot\nabla\right)\vec{u}\\ \rho\nabla\cdot\vec{u}=0\end{gathered} \tag{3.9}$$

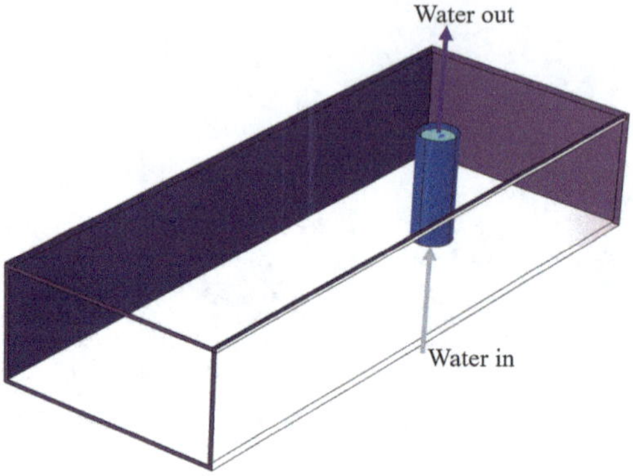

Fig. 3.5 Geometry of BJ22 waveguide with liquid

where $\vec{u}$ indicates the velocity in the open channel (m/s), $\vec{I}$ is the identity matrix, and p is the pressure (Pa).

3.3.2.3 Input Parameters

Property	Applied domains	Value
Relative dielectric constant	Air	1
Relative permittivity	Water	72-j*15
Relative permeability	Air Water	1 1
Conductivity (S/m)	Air Water	0 5.5e-6
Heat conductivity coefficient (W/m K)	Water	0.599
Density (kg/m^3)	Water	1000
Heat capacity at constant pressure (J/kg K)	Water	4186

3.3.2.4 Results

Figure 3.6 shows that the electric field in the microwave cavity is relatively concentrated near the fluid, which demonstrates an efficient utilization of microwave energy. Figure 3.7 shows that the more power is lost at the center of the fluid cylindrical. Figure 3.8 shows that after a different time of microwave heating, the fluid in the heating cavity has experienced a relatively uniform heating effect.

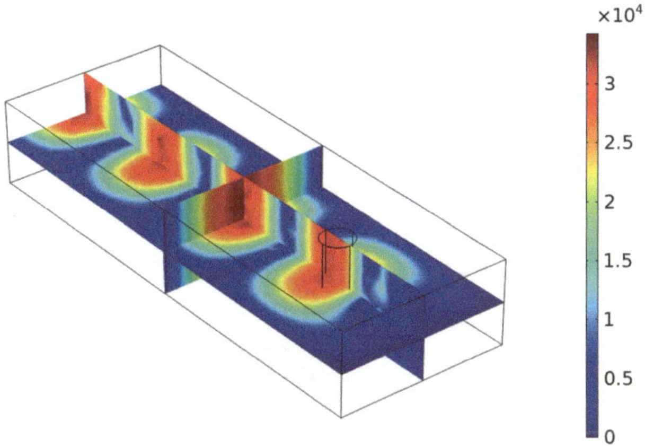

Fig. 3.6 Electric field distribution (V/m)

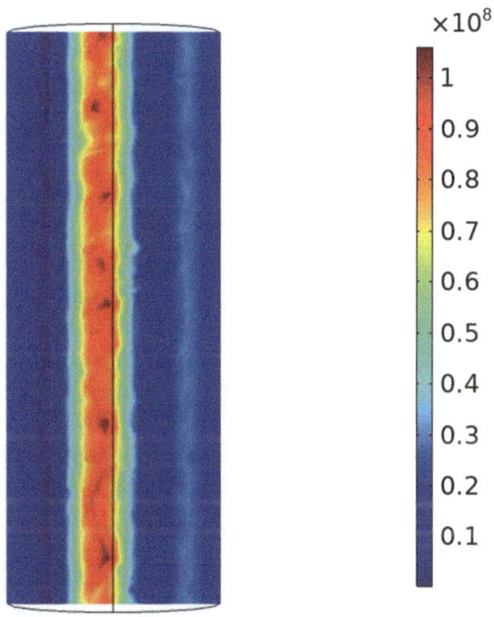

Fig. 3.7 Power loss distribution (W/m^3)

3.3.3 Transformation Optics Algorithm for Computing the Heating Process of Rotating Elements

In the microwave heating simulation, there are many problems such as excessive memory consumption and inaccurate calculations. Here, on the basis of transformation optics, a novel continuous algorithm to accurately and conveniently compute the heating process in microwave heating systems with moving elements is presented. Optical variation is used in the simulation to achieve the effect of a moving boundary

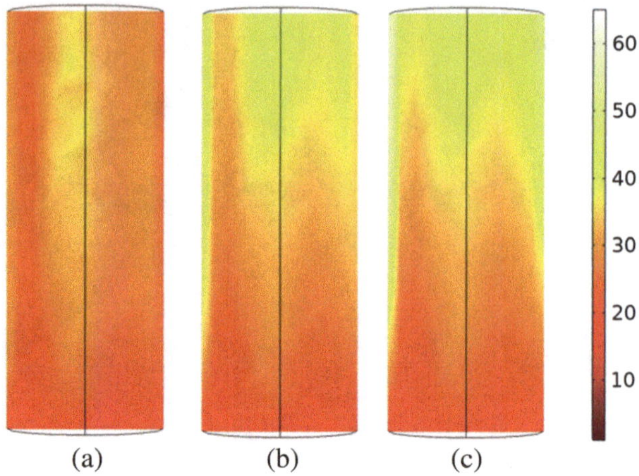

Fig. 3.8 Temperature distribution after (**a**) 2 s, (**b**) 5 s, and (**c**) 10 s of heating (degC)

and the heating process can be computed by surrounding the moving elements with time-varying permittivity and permeability.

Generally, an electric field is described by Maxwell's equations. In a time-harmonic field, Maxwell's equations can be reduced to a vector wave equation [2].

$$\nabla \times \varepsilon^{-1} \nabla \times \vec{\mathrm{E}}(r) - \omega^2 \mu \vec{\mathrm{E}}(r) = 0 \tag{3.10}$$

where $\vec{\mathrm{E}}$ is the electric field strength, ε is the permittivity, μ is the permeability, ω is the angular frequency, and r is the position vector. As mentioned above, it is difficult to mesh the dynamical locations of objects in mobile microwave heating systems. Fortunately, this problem can be resolved using transformation optics. It has been reported that this equation is form-invariance under any coordinate transformation [3–5]. In other words, the vector wave equation still preserves its form inside the transformation domain. Besides, the medium transforms into an anisotropic medium conveying the effect of coordinate transformation to the electromagnetic field.

Generally, we define a coordinate transformation as

$$r \rightarrow r' = T(r) \tag{3.11}$$

where $r = (x, y, z)$ and $r' = (x', y', z')$ are the position vectors of the original and transformed coordinates, respectively. This coordinate transformation affects both the permittivity and permeability of the surrounding region, as expressed in Eqs. (3.12)–(3.15): [4, 6].

$$\overline{\overline{\varepsilon}} = \varepsilon \overline{\overline{\gamma}} \tag{3.12}$$

$$\overline{\overline{\mu}} = \mu\overline{\overline{\gamma}} \tag{3.13}$$

where $\overline{\overline{\varepsilon}}$ and $\overline{\overline{\mu}}$ are the relative permittivity and permeability of the surrounding region after coordinate transformation. And

$$\overline{\overline{\gamma}} = \overline{J}^T \cdot \overline{J}/\det\overline{J} \tag{3.14}$$

where $\overline{J}$ is the Jacobian tensor, expressed as

$$\overline{J} = \begin{bmatrix} \frac{\partial x'}{\partial x} & \frac{\partial x'}{\partial y} & \frac{\partial x'}{\partial z} \\ \frac{\partial y'}{\partial x} & \frac{\partial y'}{\partial y} & \frac{\partial y'}{\partial z} \\ \frac{\partial z'}{\partial x} & \frac{\partial z'}{\partial y} & \frac{\partial z'}{\partial z} \end{bmatrix} \tag{3.15}$$

The field in the original and transformed coordinates can also be directly linked with the Jacobin tensor $\overline{J}$: [6].

$$E(r) = E'(r')/\overline{J}^T \tag{3.16}$$

and

$$H(r) = H'(r')/\overline{J}^T \tag{3.17}$$

Figure 3.8a illustrates a microwave applicator system with moving elements. In this system, the sample is in motion, while the electric field changes dynamically. This dynamic process indicates that the motion region remains static, while the incident electromagnetic twists with time. As such, it can be computed using transformation optics, which is illustrated schematically in Fig. 3.8b.

In this figure, the motion region, which is denoted as Ω, is surrounded by a layer of inhomogeneous and anisotropic medium. The inner and outer boundaries are denoted as $\partial\Omega_1$ and $\partial\Omega_2$, respectively. We assume that at $t = 0$, $\partial\Omega_1$ and $\partial\Omega_2$ should satisfy the following conditions:

$$f_1(r) = 0 \tag{3.18}$$

$$f_2(r) = 0 \tag{3.19}$$

where $f_1(r)$ and $f_2(r)$ are the functions used to describe geometrical boundaries. For example, in the case where $\partial\Omega_1$ and $\partial\Omega_2$ represent circles with radii of a and b, respectively, whose center points for both are (0.0), then $f_1(r)$ and $f_2(r)$ can be expressed as $f_1(r) = x^2 + y^2 - a^2$ and $f_2(r) = x^2 + y^2 - b^2$, respectively. Simultaneously, the position vector in domain Ω satisfies

$$r' = r \tag{3.20}$$

During the moving process, we assume that the object moves to a new position at $t = t_1$. The outer boundary $\partial\Omega_2$ is still consistent with the boundary condition at $t = 0$, namely

$$f_2(r) = 0 \tag{3.21}$$

However, the inner boundary $\partial\Omega_1$ changes during this process. As mentioned before, Eq. (3.9) shows the relation between r' in the transformed coordination and r' in the original coordination. Therefore, the medium parameters inside the surrounding layer can be derived from Eqs. (3.12)–(3.15). Note that the coordinate transformation $T(r)$ should change dynamically as the location of the moving elements varies with time. Namely, the permittivity and permeability inside the surrounding layer is time-dependent during this process.

3.3.4 *Implicit Function and Level Set Methods for the Computation of Moving Elements During Microwave Heating*

Although the transformation optics method possesses some unique advantages, it still cannot solve the problems associated with irregular shapes or random motions. Here, we propose a novel method to model the moving motion, which is more universal than the transformation optics method.

In the computation of the microwave heating process with moving elements, an implicit function ϕ is used to divide a closed physical domain into three parts, namely the exterior region, the interior region, and the boundary. A value of the implicit function larger than 0 refers to the exterior region. On the contrary, the value equivalent to 0 and that below 0 refer to the boundary and the interior region, respectively. The interior and exterior regions of the physical domain can be easily demonstrated using a Heaviside function (Fig. 3.9):

$$H(\varphi) = \begin{cases} 1, & \varphi > 0 \\ 0, & \varphi < 0 \end{cases} \tag{3.22}$$

where φ depends on the coordinates of the points. Therefore, whether a point is inside or outside of the graph is indicated by the value of $H(\varphi)$.

However, since implicit functions cannot represent the movement of the elements in the microwave heating process, level set methods are introduced to deal with this problem. The implicit function φ employed in the level set methods is defined as a signed distance function [7]:

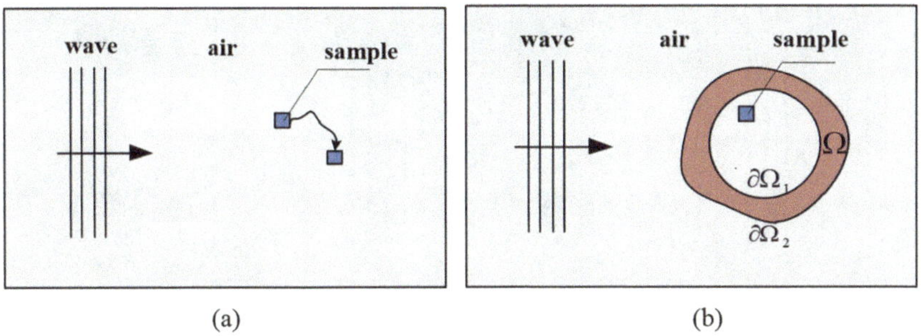

Fig. 3.9 Sketch of a microwave applicator system with moving elements. (**a**) Realistic motion and (**b**) computed by surrounding the motion region with a layer of time-varying medium based on transformation optics

$$|\varphi(r)| = d(r) = \min_{r_I \in \partial\Omega}(|r - r_I|) \tag{3.23}$$

where $\partial\Omega$ is the interface; $\phi(r) > 0$ refers to the exterior; $\phi(r) < 0$ refers to the interior; and $\phi(r) = 0$ refers to the boundary, as mentioned previously.

Provided that the velocity of the element is $V(r)$, in the simulation, the movement of the element can be expressed by the following simple ordinary differential equation [7]:

$$d(r_I)/dt = V(r_I) \tag{3.24}$$

where r_I denotes the position vector of every point on the boundary of the element. Then, the level set Eq. (3.25) is used to define the evolution of the implicit function according to Eq. (3.24) [7]:

$$\frac{d\varphi}{dt} + V \cdot \nabla\varphi = 0 \tag{3.25}$$

Eq. (3.25) indicates that the implicit function of the interface of the next time step is related to that of the last time step. Here, we do not consider the shape change of the elements during microwave heating. Thus, with Eqs. (3.23) and (3.24), we can obtain the implicit function by updating the position of the points on the boundary of the element after each time step.

In the practical calculation, the elements in the model can be considered as a combination of a series of closed physical domains. As illustrated in Fig. 3.10, an element is rotating under exposure to a microwave with a velocity of V(r). The solid line region represents the boundary of the element at time t_n, while the dashed line region represents the boundary of the element at the next time step with $t_{n+1} = t_n + \Delta t$. Assuming that the implicit function of the boundary of the element is φ^n at time t_n, then the mesh dots inside or outside of the element in the moving region can be

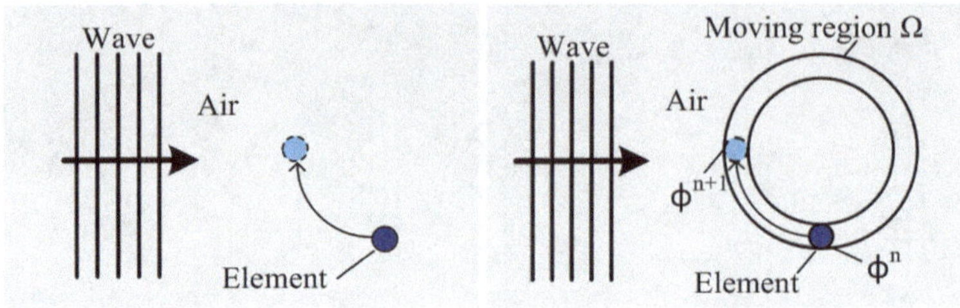

Fig. 3.10 Sketch of a microwave applicator system with a moving element in a hypothetical case. (**a**) Realistic movement. (**b**) Computed by adding a moving region represented by an implicit function

recognized by the value of $H(\varphi^n)$, as discussed in Eq. (3.22). Therefore, the complex permittivity of the moving region at time t_n can be defined as

$$\varepsilon_m = \varepsilon_1[1 - H(\varphi^n)] + \varepsilon_2 H(\varphi^n) \tag{3.26}$$

where ε_1 denotes the relative permittivity of the element and ε_2 denotes the permittivity of air. The permeability of the moving region can be expressed in the same form. Generally, the permeability of the whole region is regarded to be the same in most cases. According to Eq. (3.24), there are some mesh elements near the boundary of the element with some nodes satisfying $H(\varphi^n) = 0$ and others satisfying $H(\varphi^n) = 1$. Inside these mesh elements, $H(\varphi^n)$ is obtained by interpolating the $H(\varphi^n)$ of their nodes, which will cause errors in the calculation. This error is similar to the inherent one of FEM, which can be reduced by refining the mesh size. Therefore, a mesh-independent study is needed in the calculation.

At time t_{n+1}, the element moves to position $n + 1$. To update the position of the element, the forward Euler method is used for the time discretization of Eq. (3.23). Therefore, the position vector of the points on the boundary of the element can be expressed as

$$r_I^{n+1} = r_I^n + V\left(r_I^n\right) \cdot \Delta t \tag{3.27}$$

where r_I^n is the position vector of the points on the boundary of the element at time t_n. Then, by using Eqs. (3.23) and (3.25), φ^{n+1} can be updated as

$$\left|\varphi^{n+1}\left(r^{n+1}\right)\right| = d\left(r^{n+1}\right) = \min_{r_I \in \partial\Omega}\left(\left|r - r_I^{n+1}\right|\right) \tag{3.28}$$

During the heating process, the complex permittivity of the moving region is also updated repeatedly after each time step by using Eqs. (3.25)–(3.28). Therefore, the electric field distribution can be obtained by solving Maxwell's waveform, Eq. (3.29), with a time-varying complex permittivity [8]:

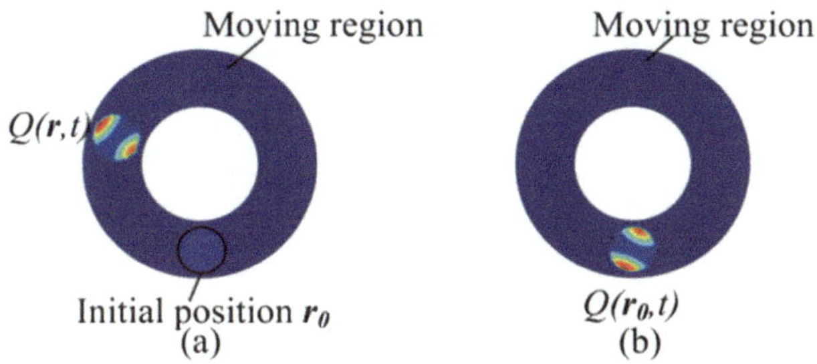

Fig. 3.11 Sketch of the dissipated power of the electromagnetic field in the moving region in a hypothetical case. (**a**) Variation of the dissipated power with time in the original calculation. (**b**) Variation of the dissipated power with time after transformation

$$\nabla \times \mu_r^{-1}\left(\nabla \times \vec{E}\right) - k_0^2\left(\varepsilon_r - j\frac{\sigma}{\omega\varepsilon_0}\right)\vec{E} = 0 \tag{3.29}$$

where $\vec{E}$ is the electric field, ε_0 is the permittivity of vacuum, ε_r is the relative permittivity of the medium, μ_r denotes the relative permeability, ω represents the angular frequency, k_0 is the wave number in vacuum, and σ is the electrical conductivity.

For computing the heating process, the dissipated power of the electromagnetic field in the moving region can be obtained from Eq. (3.3) [9].

According to Eqs. (3.26)–(3.28), the coordinates of the dissipated power in the moving region change with the realistic movement of the element, as shown in Fig. 3.11a, where the dissipated power distribution of the element is obtained for a hypothetical case. However, to calculate the temperature distribution of the element, some additional settings are required because it is difficult for the heat conduction equation to follow the movement of the element. Fortunately, this problem can be resolved by transforming the heat source back to the initial position of the element before performing the heat transfer calculation in each time step, which keeps the coordinates of the heated position of the element in the calculation process, as shown in Fig. 3.11b. To make the coordinates after transformation consistent with the initial coordinates, the effect of coordinate transformation and that of the movement equation of the element should be inversed. For example, when the velocity of the element is $V(r)$, the transformation equation in each time step can be expressed as

$$r_0 = r^n - \sum_{m=1}^{n} V(r^m) \cdot (m-1)\Delta t \tag{3.30}$$

where r_0 is the initial position vector of the points of the element and r^n is the position vector of the points of the element at the nth time step. With Eq. (3.30), we only need to model the initial position of the element in the simulation because the heat source can be transformed to the initial position perfectly by extrapolation.

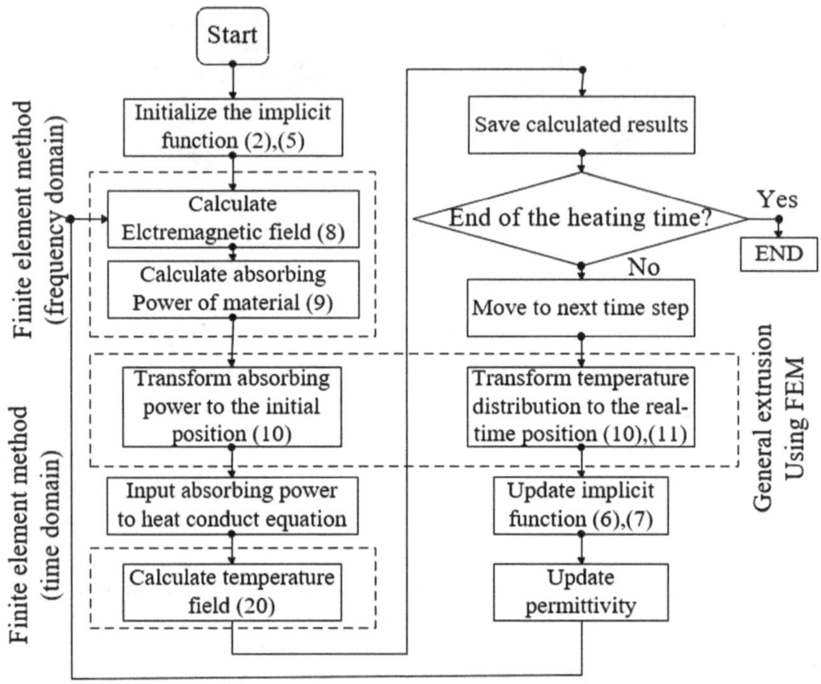

Fig. 3.12 Overall flowchart

To update the complex permittivity of the moving element according to temperature, the temperature distribution needs to be transformed into the realistic position of the moving element after each time step, which can be expressed as

$$T(\mathrm{r}_0, t) \rightarrow T_p^n(\mathrm{r}^n, t) \tag{3.31}$$

where T_p^n is the temperature after transformation at the nth time step, which is used to update the complex permittivity of the moving region. The overall flowchart is shown in Fig. 3.12. The proposed method can also be applied to the calculation of microwave heating with mode stirrers. In this case, we only need to use Eqs. (3.26)–(3.29) to calculate and update the electromagnetic field distribution without transforming the heat source and temperature distribution.

3.4 Microwave Heating of Chemical Reactions

3.4.1 Introduction

A chemical reaction system is dynamic and time-varying, which is hard to analyze, but it is important for industrial applications. Therefore, in this chapter, we take a fast reaction as an example and use dielectric characterization to study the specificity of microwave propagation in chemical reactions. Especially, the reactions of acetone iodide and biodiesel are studied to show the heating effects of microwaves. The representation of the reaction polarization is reduced to that of the dielectric coefficient.

3.4.2 Multiphysics Simulation of Microwave Propagation in a Fast Reaction

The propagation of microwaves in general dispersive media has been studied extensively for decades. To the best of our knowledge, studies of microwave propagation characteristics in chemical reactions have rarely been reported. The reason for this might be that previous studies on electrodynamics in chemical reactions are mainly based on the local equilibrium theory, which can mostly be expressed by concentration and temperature. However, most chemical reactions are nonequilibrium processes. The expressions of electrodynamics face great difficulties in chemical reactions, which are important for practical applications and academic interest. Furthermore, it is inferred that the most fundamental differences between general dispersive media and chemical reactions is attributed to the fact that the former have a constant dispersive characteristic, while the latter have a time-varying dispersive characteristic.

Generally, the number of dipoles changes as the chemical reaction proceeds. The distribution functions of the reactants and products may change in chemical reactions. The polarization of chemical reactions under the exposure to microwaves can be obtained by summing the polarization of the reactant molecules and the product molecules per unit volume. To make the expressions of electrodynamics easy, without loss of generality, a simple polar-molecule reaction is chosen, for which there are special and representative model reactions. We consider the effect of polarization of only one component X_i on the microwave propagation during the reaction. X_i can be the reactant or the product. In reality, the simple polar-molecule reactions can be encountered in practical applications, such as isomerization reactions and unimolecular reactions.

Based on the theory of molecule reorientation, the expression of polarization is derived using a modified Smoluchowski-Debye equation. The Legendre polynomials are applied to express the distribution functions of the reactants and products. To use low-field approximation, the distribution functions of the reactant and product are mainly decided by the Legendre polynomials of the leading order and first order. Combined with the chemical reaction kinetics equation for the reactions, the polarization of Xi is given as follows:

$$P(t) = \phi_{X_i}(t) * [C_{X_i}(t)E(t)] \tag{3.32}$$

Here, $*$ denotes the convolution. $\phi_{X_i}(t)$ represents the rotational diffusion properties of X_i, while $C_{X_i}(t)$ denotes the concentrations of X_i, where $\phi_{X_i}(t) = \varepsilon_0 \chi_{X_i} \exp(-t/\tau_{X_i})/\tau_{X_i}$, χ_{X_i} is the static electric susceptibility, ε_0 is the vacuum permittivity, and τ_{X_i} is the dielectric relaxation time. Moreover, χ_{X_i} can be indicated in the form $\chi_{X_i} = \frac{N_A \mu_{X_i}^2}{3\varepsilon_0 K_B T}$, where μ_{X_i}, N_A, K_B, and T denote the magnitude of the dipole moment, Avogadro constant, Boltzmann constant, and the temperature of the reaction, respectively.

Generally, the polarization of a simple polar-molecule reaction might be affected by the dielectric properties of X_i, the reaction rate, and the initial concentration of the reactant. The parameters of reaction rate and initial reactant concentration are fundamental properties of chemical reactions. Therefore, the effects of reaction rate and initial reactant concentration on the characteristics of the microwaves should be investigated by understanding the equations of electromagnetic wave propagation deeply.

1. The attenuation and dispersion characteristics of microwave propagation in simple polar-molecule reactions

The spatiotemporal envelopes of microwaves are subjected to complex variations when they pass through nonstationary media. A simple polar-molecule reaction can be regarded as a nonstationary medium. The attenuation and dispersion characteristics of microwave propagation in the reactions would change with time, reaction rate, and initial reactant concentration. To disclose the influences of reaction rate and initial reactant concentration on the attenuation and dispersion characteristics, we select an instant in the reaction. The microwave propagation in the reaction is characterized by the electric flux density, $\vec{\mathrm{D}}$ and magnetic flux density, $\vec{\mathrm{B}}$, which depend on the electric field, $\vec{\mathrm{E}}$, and magnetic field, $\vec{\mathrm{H}}$, as $\vec{\mathrm{D}} = \varepsilon_0 \vec{\mathrm{E}} + \vec{\mathrm{P}}$ and $\vec{\mathrm{B}} = \mu_0 \vec{\mathrm{H}}$. In addition, ε_0 and μ_0 are the vacuum permittivity and permeability, respectively, while $\vec{\mathrm{P}}$, shown in Eq. (3.32), is the polarization of the reaction. The reaction is assumed to be a homogeneous isotropic nonmagnetic medium, being without free charges and currents. It is straightforward to obtain the following propagation equations.

$$\begin{aligned}\nabla \times \nabla \times \vec{E} = -\nabla^2 \vec{E} + \nabla\left(\nabla \cdot \vec{E}\right) = -\mu_0 \varepsilon_0 \frac{\partial^2 \vec{E}}{\partial t^2} - \mu_0 \frac{\partial^2 \vec{P}}{\partial t^2} \\ \nabla \times \nabla \times \vec{H} = -\nabla^2 \vec{H} + \nabla\left(\nabla \cdot \vec{H}\right) = -\mu_0 \varepsilon_0 \frac{\partial^2 \vec{H}}{\partial t^2} + \frac{\partial\left(\nabla \times \vec{P}\right)}{\partial t}\end{aligned} \tag{3.33}$$

As $\nabla\left(\nabla \cdot \vec{E}\right)$ is typically much smaller than $\nabla^2 \vec{E}$ and the electric and magnetic fields have the relation $\nabla \times \vec{E} = -\partial \vec{B}/\partial t$, we just consider the electric field equation, which is written as

$$\nabla^2 \vec{E} = \mu_0 \varepsilon_0 \frac{\partial^2 \vec{E}}{\partial t^2} + \mu_0 \frac{\partial^2 \vec{P}}{\partial t^2} \tag{3.34}$$

The reaction may affect the frequency of the incident microwaves. When the incident field is supposed as a monochromatic plane wave with angular frequency, ω_0, and the reaction rate is much lower than ω_0, $\vec{E}$ is assumed as a quasi-monochromatic wave propagating along the z-direction with ω_0. We use β_0 to represent the phase constant of the electromagnetic wave propagation at the instant t_0. That is, $\vec{E} = \vec{e_x} E \exp\left[i(\beta_0 z - \omega_0 t)\right] + c.c.$, where $c.c.$ denotes the complex conjugate. The envelope $\vec{E}$ of the electric field changes slowly in t and z.

The inverse Fourier transformation is carried out on the polarization, so Eq. (3.32) can be rewritten as

$$\vec{P}(t) = \frac{1}{2\pi} \int_{-\infty}^{+\infty} P(\omega) \exp(-i\omega t) d\omega = \varepsilon_0 \frac{1}{2\pi} \int_{-\infty}^{+\infty} \chi_{X_i} E_{C_{X_i}}(\omega) \exp(-i\omega t) d\omega \tag{3.35}$$

where $\chi_{X_i}(\omega) = \int_{-\infty}^{+\infty} \chi_{X_i}(t) \exp(i\omega t) dt$ and $E_{C_{X_i}}(\omega) = \int_{-\infty}^{+\infty} \left[C_{X_i}(t) \vec{E}(t)\right] \exp(i\omega t) dt.$

We expand $\chi_{X_i}(\omega)$ at ω_0 in the Taylor series as

$$\chi_{X_i}(\omega) = \sum_{n=0}^{\infty} \frac{1}{n!} \frac{\partial^n \chi_{X_i}(\omega)}{\partial \omega^n}\bigg|_{\omega=\omega_0} (\omega - \omega_0)^n \tag{3.36}$$

Subsequently, substituting Eq. (3.36) into Eq. (3.35) and using the shift $\omega \to \omega - \omega_0$, Eq. (3.35) can be written as

$$\vec{\mathrm{P}}(t) = \varepsilon_0 \sum_{n=0}^{\infty} \frac{i^n}{n!} \exp(-i\omega_0 t) \times \left\{ \frac{\partial^n \chi_{X_i}(\omega)}{\partial \omega^n}\bigg|_{\omega=\omega_0} \frac{\partial^n \left[C_{X_i}(t)\vec{\mathrm{E}}(t) \exp(i\omega_0 t) \right]}{\partial t^n} \right\} \tag{3.47}$$

Now, when the reaction is supposed as a low-loss medium, we obtain the following electric field envelope equation.

$$\begin{aligned} \frac{\partial E}{\partial z} &= -i\frac{\beta_0}{2}E + i\frac{\mu_0\varepsilon_0}{2\beta_0}\omega_0^2 E - \frac{\mu_0\varepsilon_0}{\beta_0}\omega_0\frac{\partial E}{\partial t} - i\frac{\mu_0\varepsilon_0}{2\beta_0}\frac{\partial^2 E}{\partial t^2} \\ &+ i\frac{\mu_0}{2\beta_0}\mathrm{M}_{0r}E - \frac{\mu_0}{2\beta_0}\mathrm{M}_{0i}E - \frac{\mu_0}{2\beta_0}\mathrm{M}_{1r}\frac{\partial E}{\partial t} - i\frac{\mu_0}{2\beta_0}\mathrm{M}_{1i}\frac{\partial E}{\partial t}\cdots \end{aligned} \tag{3.38}$$

where

$$\mathrm{M}_0 = R_0 C_{X_i}(t) + iR_1\frac{\partial C_{X_i}(t)}{\partial t} - R_2\frac{\partial^2 C_{X_i}(t)}{\partial t^2} + O\left(\frac{\partial^2}{\partial t^2}\right)$$

$$\mathrm{M}_{0r} = \mathrm{Re}(\mathrm{M}_0) \quad \mathrm{M}_{0i} = \mathrm{Im}(\mathrm{M}_0) \ \ \mathrm{M}_1 = R_1 C_{X_i}(t) + i2R_2\frac{\partial C_{X_i}(t)}{\partial t} + O\left(\frac{\partial}{\partial t}\right)$$

$$\mathrm{M}_{1r} = \mathrm{Re}(\mathrm{M}_1) \quad \mathrm{M}_{1i} = \mathrm{Im}(\mathrm{M}_1)$$

with

$$R_0 = \omega^2\phi_{X_i}(\omega)\big|_{\omega=\omega_0} \quad R_1 = \left[2\omega\phi_{X_i}(\omega) + \omega^2\frac{\partial\phi_{X_i}(\omega)}{\partial\omega}\right]\bigg|_{\omega=\omega_0}$$

$$R_2 = \left[\phi_{X_i}(\omega) + 2\omega\frac{\partial\phi_{X_i}(\omega)}{\partial\omega} + \frac{\omega^2}{2}\frac{\partial^2\phi_{X_i}(\omega)}{\partial\omega^2}\right]\bigg|_{\omega=\omega_0}$$

The form of Eq. (3.38) suggests that at t_0, the attenuation characteristic is

$$\alpha = \frac{\mu_0}{2\beta_0}\mathrm{M}_{0i} = \omega_0\sqrt{\mu_0\varepsilon_0}\frac{\mathrm{M}_{0i}/\left(\omega_0^2\varepsilon_0\right)}{2\sqrt{1 + \mathrm{M}_{0r}/\left(\omega_0^2\varepsilon_0\right)}} \tag{3.39}$$

and the phase characteristic is

$$\beta_0 = \omega_0 \sqrt{\mu_0 \varepsilon_0 \left(1 + \frac{M_{0r}}{\varepsilon_0 \omega_0^2}\right)} \tag{3.40}$$

In fact, when the rate constant is zero, which means that the reaction can be considered as a mixture of the reactants, $C_{X_i}(t) = C_{X_i}^0$ is obtained subsequently. Thus, the attenuation and phase characteristics can be simply expressed as

$$\alpha = \omega \sqrt{\mu_0 \varepsilon_0} \frac{\mathrm{Im}\left[\phi_{X_i}(\omega) C_{X_i}^0\right] / \varepsilon_0}{2\sqrt{1 + \mathrm{Re}\left[\phi_{X_i}(\omega) C_{X_i}^0\right] / \varepsilon_0}} \tag{3.41}$$

$$\beta_0 = \omega \sqrt{\mu_0 \varepsilon_0 \left\{1 + \mathrm{Re}\left[\phi_{X_i}(\omega) C_{X_i}^0\right] / \varepsilon_0\right\}} \tag{3.42}$$

We can recover the same formulas of the attenuation and phase characteristics in low-loss dispersive media. Therefore, the attenuation and dispersion characteristics obtained above are more generic than the ones in a dispersive medium. Moreover, the role of $C_{X_i}(t)$ in the microwave propagation is demonstrated in that it modulates the attenuation and dispersion characteristics remarkably for the given conditions.

To further validate the derived approximate expressions, comparisons are made between the numerical solutions achieved by the FETD method and the approximate values obtained from Eqs. (3.39) and (3.40). We calculate the propagation of the monochromatic plane wave in the reactions using the 1D mold. The traverse electromagnetic plane wave is normally incident at the interface between air and where the reaction takes place.

The reaction rate and initial reactant concentration are key features in the reaction, which affect α and β_0 differently. When the incident power is low, the temperature change of the reaction can be neglected. Therefore, $\chi_{X_i} C_{X_i}^0$ can be considered to be variable with the initial reactant concentration. Furthermore, as the reaction can be considered as a nonstationary medium, α and β_0 would change with time. Thus, to know the influence of the reaction rate and initial reactant concentration on α and β_0, two conditions are considered at t_0. The parameters used for the calculation are listed in Table 3.1, where $\overline{k}$ is the relation of the forward rate k1 and the reverse reaction. Combining this with the chemical reaction kinetics equation of the first-order

Table 3.1 Parameters for the reactions

	$\overline{k}$	$\tau_{X_i}(s)$	$k_1(\mu s^{-1})$	$\chi_{X_i} C_{X_i}^0$
Condition I	0.5	1×10^{-11}	0, 0.01, 0.1, 1, 2, 3, 4, 5, 6, 7, 8, 9, 10	10
Condition II	0.5	1×10^{-11}	10	0, 10, 20, 30, 40, 50, 60, 70, 80, 90, 100

Table 3.2 Results of the parameters

τ_{A_1}(s)	μ_{A_1}(D)	$c_{A_1}(0)$(mol/L)	T (K)	k (s-1)
1×10^{-12}	0.136	10	300	1×10^3, 1×10^8, 1×10^{10}

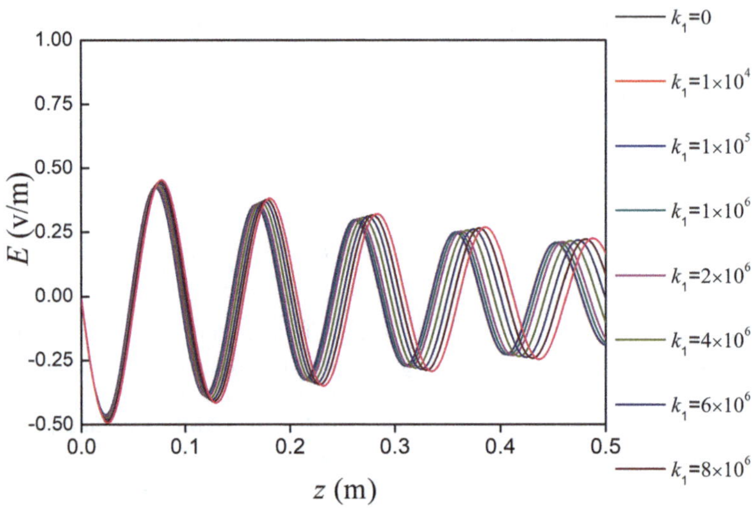

Fig. 3.13 Electric field distributions along the propagation direction with the variation of k1 under Condition I

reaction, $C_{X_i}(t) = C^0_{X_i}\left\{\frac{\bar{k}}{1+\bar{k}} + \frac{\exp\left[-\left(1+\bar{k}\right)k_1 t\right]}{1+\bar{k}}\right\}$ can be obtained when X_i is the reactant.

Here, the incident electromagnetic wave is set as follows:

$$E(t) = \sin(2\pi f_0 t) \tag{3.43}$$

where f_0 is 1 GHz. The mesh size and time step are 3 mm and 10 ps, respectively. The electrical field distribution is computed from the interface to 0.5 m beneath the interface.

Under Condition I, Fig. 3.13 shows the electric field distributions along the propagation direction at $t_0 = 20$ ns with the variation of k1 using the FETD method. The results indicate that the amplitude and wavelength of the electric field rise gradually with an increase in k_1.

Variations of α and β_0 under different values of k_1 obtained by the FETD method are shown in Fig. 3.14, along with the values obtained from the approximate expressions. The deviations between the FETD method and the approximate expressions are also shown in Fig. 3.14. A higher concentration of X_2 can be obtained under a faster k_1 when $\chi_{X_i}C^0_{X_i}$ is fixed, which simultaneously indicates a lower

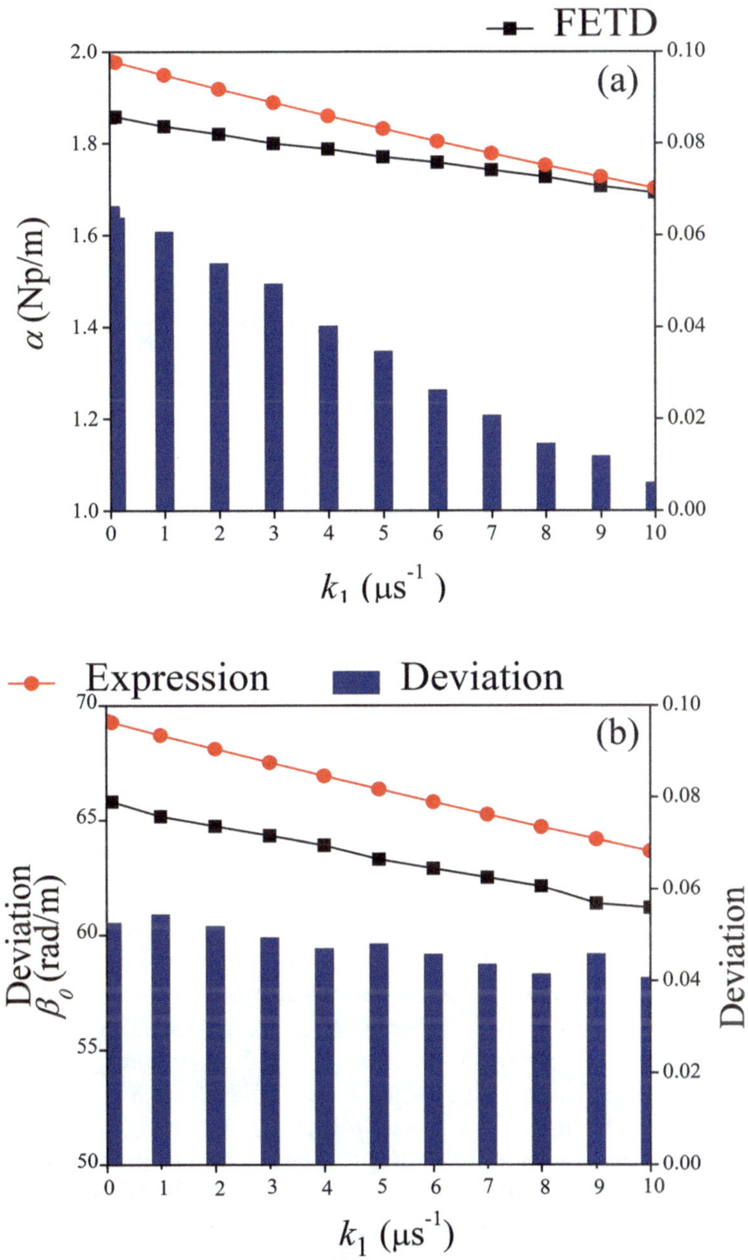

Fig. 3.14 Evolution of α and β_0 with variation of k1 under Condition I. (**a**) α and (**b**) β_0

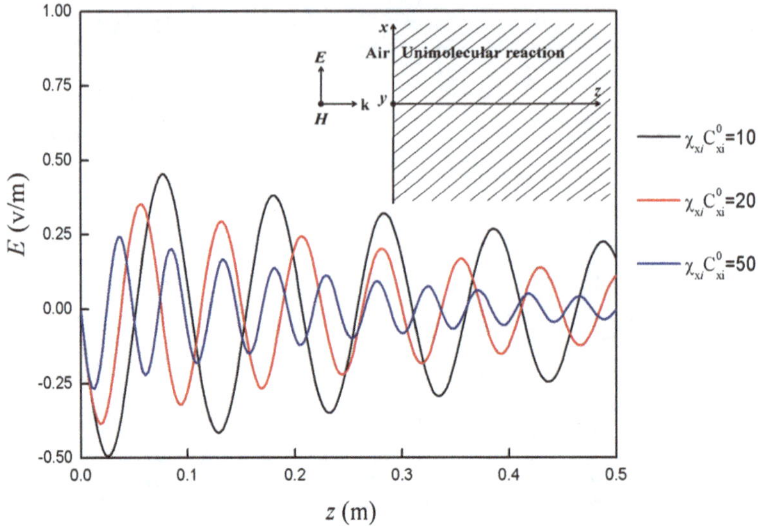

Fig. 3.15 Electric field distributions along the propagation direction at different $\chi_{X_i}C^0_{X_i}$ under Condition II

concentration of X_1. Furthermore, it is assumed that X_I has a permanent dipole moment, while X_2 has none. Therefore, the quantity of dipoles in the reaction is affected solely by the variation of X_1, which would be smaller at a higher k_1. It can be inferred that the impedance of the reaction on the microwave propagation is weakened by the decrease in the number of dipoles. Thus, the increase in k_1 leads to reductions in α and β_0. In addition, the average deviations between the FETD method and the approximate expressions for α and β_0 are 4.8% and 4%, respectively.

Figure 3.15 shows the electric field distributions along the propagation direction under Condition II, using the FETD method at $t_0 = 20$ ns. In Fig. 3.15, the amplitude and wavelength of the electric field decrease with an increase in $\chi_{X_i}C^0_{X_i}$.

The parameters of α and β_0 under different values of $\chi_{X_i}C^0_{X_i}$ are obtained by the FETD method and the approximate expressions are presented in Fig. 3.16. The improvement in $\chi_{X_i}C^0_{X_i}$ leads to an increase in α and β_0, and more specifically, with an increase in the initial concentration of X_I, which can also be explained by the different number of dipoles. Moreover, α and β_O are matched within average deviations of 2.5% and 2.9%, respectively. Thus, with the variations of k_I and $\chi_{X_i}C^0_{X_i}$, the relatively small deviations between the FETD method and the approximate expressions indicate that the approximate expressions can state the characteristics of microwave propagation well.

In summary, we derive the envelope equation of microwave propagation in the reactions. The approximate expressions on the attenuation and dispersion characteristics of microwave propagation are obtained. The results achieved by the FETD method and by the approximate expressions are compared, and can validate the feasibility of the approximate expressions. The values of the attenuation and

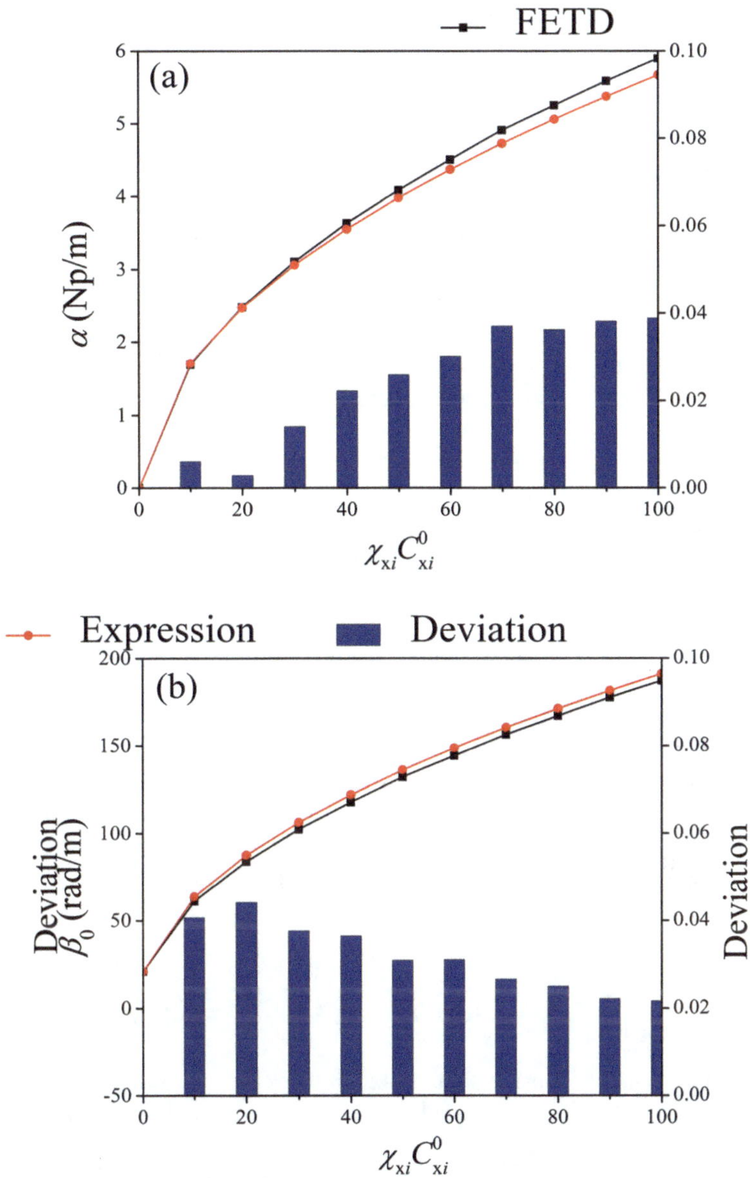

Fig. 3.16 Variations of α and β_0 at different $\chi_{X_i} C_{X_i}^0$ under Condition II. (**a**) α and (**b**) β_0

dispersion characteristics decrease gradually with an increase in the reaction rate and an increase in the enhancement of the initial reactant concentration. These trends can be explained by the change in the number of dipoles. These results may be helpful in disclosing the characteristics of electromagnetic wave propagation in a chemical reaction.

2. Frequency changes of microwaves in simple polar-molecule reactions

Most chemical reactions are nonequilibrium processes, where changes in the status affect the characteristics of microwaves in the chemical reactions. To disclose the propagation law of microwaves in nonequilibrium processes, Morgenthaler revealed the frequency transformations in unbounded media whose permittivity changes with time. Kalluri discussed frequency changes in time-varying plasma. For chemical reactions treated with microwaves, the transformation of the frequency would be a reasonable method to unfold the effect of the chemical reaction's status on microwaves. In this section, based on the theoretical formula of polarization in a simple polar-molecule reaction, we transform the propagation equation of microwaves in the reaction to achieve the electromagnetic properties of the reaction. By comparing the propagation equation of microwaves in general media, the simple polar-molecule reactions can be described by the rotational diffusion vector and the component concentration vector, which are proposed to describe the polarization. Subsequently, using the finite-element time domain (FETD) method, frequency transformations are investigated according to the variation in the component concentration in the reaction.

The electromagnetic properties of the media are reflected in mathematical properties of the response functions. For simple media, the electromagnetic properties can be exhibited by anisotropy or dispersion. However, the reactions adopted in the current research are complex media because the time-varying parameter $C_{X_i}(t)$ is exhibited in the function, as shown in Eq. (3.32). To know the electromagnetic properties of the reactions mentioned in Eq. (3.32), the reactions are assumed as homogeneous isotropic nonmagnetic media, meaning they are without free charges and currents. The microwaves in this type of chemical reaction are governed by Maxwell's equations. The electric field equation can be written as

$$\nabla \times \nabla \times \vec{E} = -\mu_0 \frac{\partial^2 \vec{D}}{\partial t^2} = -\mu_0 \varepsilon_0 \frac{\partial^2 \vec{E}}{\partial t^2} - \mu_0 \frac{\partial^2 \vec{P}}{\partial t^2} \tag{3.44}$$

$\vec{P}$, shown in Eq. (3.35), is the polarization of the reactions. Substituting Eq. (3.32) into Eq. (3.44), Eq. (3.45) is achieved.

$$\nabla \times \nabla \times \vec{E} = -\mu_0 \left\{ \varepsilon_0 \frac{\partial^2 \vec{E}}{\partial t^2} + \frac{\partial^2 \phi_{X_i}(t) * \left[C_{X_i}(t) \vec{E} \right]}{\partial t^2} \right\} \tag{3.45}$$

Subsequently, $C_{X_i}(t)\vec{E}$ is defined as $\vec{E}_{X_i}$, which indicates the modulated electric field in the reactions. Eq. (3.45) can be transformed into the following equation.

$$\frac{1}{C_{X_i}(t)}\nabla\times\nabla\times\vec{E}_{X_i}=-\mu_0\left\{\varepsilon_0\frac{\partial^2}{\partial t^2}\left[\frac{1}{C_{X_i}(t)}\vec{E}_{X_i}\right]+\frac{\partial^2}{\partial t^2}\left[\phi_{X_i}(t)*\vec{E}_{X_i}\right]\right\} \quad (3.46)$$

By introducing a new coordinate ($\nabla'=\frac{1}{\sqrt{C_{X_i}(t)}}\nabla$) in Eq. (3.49), Eq. (3.50) can be obtained.

$$\nabla'\times\nabla'\times\vec{E}_{X_i}=-\mu_0\left\{\varepsilon_0\frac{\partial^2}{\partial t^2}\left[\frac{1}{C_{X_i}(t)}\vec{E}_{X_i}\right]+\frac{\partial^2}{\partial t^2}\left[\phi_{X_i}(t)*\vec{E}_{X_i}\right]\right\} \quad (3.47)$$

Comparing Eq. (3.50) with Eq. (3.44), the form of Eq. (3.47) suggests that the electric displacement vector can be expressed as Eq. (3.51) with new coordinates.

$$\vec{D}_{X_i}=\varepsilon_0\frac{1}{C_{X_i}(t)}\vec{E}_{X_i}+\phi_{X_i}(t)*\vec{E}_{X_i} \quad (3.48)$$

Therefore, under the given conditions, the electromagnetic properties of the reactions exhibit dispersive and time-varying characteristics based on the mathematical properties of Eq. (3.48).

As discussed above, the main difference between the reactions mentioned in Eq. (3.32) and in general dispersive media is the time-varying characteristic, which originates from the change in the component concentrations. The parametric phenomena of microwaves in active media can be realized by a time modulation of the media parameters. When microwaves propagate in a reaction, they undergo transformations that depend on both the variations of the components in the reactions and the initial characteristics of the signal. The frequency transformations of microwaves caused by the reactions can be realized by a time modulation of the media parameters, which would be caused by the change in the component concentrations in the reactions. Moreover, frequency shifts would impact the expression method for the dielectric behavior of the reactions.

To investigate the frequency transformations in the reactions, without loss of generality, we just consider the first-order reaction with the same initial concentration ($C^0_{X_i}$), where k is the rate constant. The relation $C_{X_i}(t)=C^0_{X_i}\exp(-kt)$ can be obtained if X_i is regarded as the reactants, and the expression $C_{X_i}(t)=C^0_{X_i}(1-\exp(-kt))$ can be achieved when X_i is considered as the products.

Subsequently, numerical calculations based on the FETD method are carried out to achieve the frequency changes of microwaves in the reactions, so as to understand the effect of the reactions on microwaves. In the numerical calculations, we assume that the reactions occur at t_0. The incident plane harmonic wave is set as

$$E(t)=\sin(2\pi f_0 t) \quad (3.49)$$

where f_0 is 1 GHz. The mesh size and time step are 3 mm and 10 ps, respectively.

Furthermore, the microwave propagation in homogenous nonmagnetic reactions without free charges and currents is governed by Eq. (3.44). Therefore, the evolution of $\vec{E}(t)$ can be affected by the variation in $C_{X_i}(t)$. More specially, if the variation rate of $C_{X_i}(t)$ changes much more slowly than that of $\vec{E}(t)$, adiabatic approximations can be used to yield a slow change in the results, without transients. When the variation rate of $C_{X_i}(t)$ changes rapidly with time, transients become apparent. When X_i is regarded as the reactant, frequency changes are shown in Fig. 3.17a. For the case of $k = 1 \times 10^3\ \mathrm{s}^{-1}$, the variation rate of the reaction parameters is much slower than the rate of change of the microwaves. Thus, the media are non-transient and the frequency changes can be neglected. The effect of the non-transient condition is to merely induce the amplitude attenuation of the microwaves. When $k = 1 \times 10^8\ \mathrm{s}^{-1}$, the media parameters of the reactions undergo a smooth change with time, and to study the transformations of the microwaves in the reaction is reasonable for practical applications. The results of the frequency changes under smooth conditions are complex. For an initial harmonic wave input into the reactions, the output wave has a non-monchromatic characteristic. For $k = 1 \times 10^{10}\ \mathrm{s}^{-1}$, the variation rate of the reaction parameters is much higher than the rate of change of the microwaves, indicating that the reaction has finished immediately. Transients become apparent and the corresponding frequency shifts from 1 to 3.05 GHz.

When X_i is the product, the frequency changes are shown in Fig. 3.17b. Frequency changes are not observed as $k = 1 \times 10^3\ \mathrm{s}^{-1}$. With $k = 1 \times 10^8\ \mathrm{s}^{-1}$, frequency transformations appear in a complex way, and are smaller than the incident frequency. When $k = 1 \times 10^{10}\ \mathrm{s}^{-1}$, the frequency shifts from 1 to 0.33 GHz.

A temporal discontinuity of the dielectric properties is created as the component concentration varies. In case of temporal discontinuity, the wave number is conserved and the frequency is inversely proportional to the dielectric property. When X_i is the reactant, the component concentration of X_i decreases gradually in the reaction. It can be inferred that the dielectric property is weakened with the decrease in the number of dipoles. Thus, when X_i is the reactant, the reduction in the dielectric property leads to an increase in frequency. Moreover, when X_i is the product, the decrease in frequency can also be explained by different numbers of dipoles.

3. Relation between the time and frequency domains for the presentation of polarization in simple polar-molecule reactions

The effective permittivity of chemical reactions is proposed for slow chemical reactions, and the rotational diffusion vector and component concentration vector can be used to theoretically describe the polarization of polar-molecule reactions. Thus, under slow chemical reactions that are non-transient, the results of the unchanged temporal shape and frequency are used to achieve the relation between the effective permittivity and polarization model. An initial plane harmonic wave $\vec{E}(z,t) = \hat{e}_x \exp[i(\omega_0 t - k_0 z)]$ propagates in homogenous nonmagnetic reactions without free charges and currents. k_0 is the wave number and ω_0 is the angular

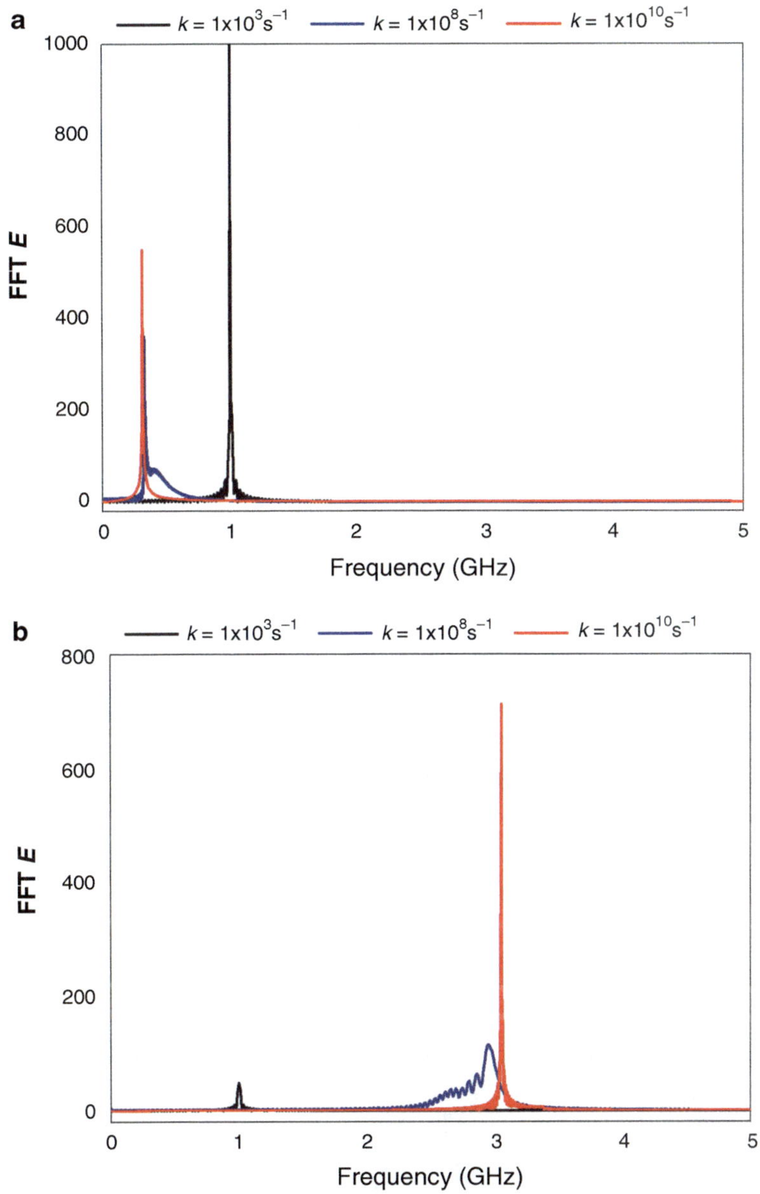

Fig. 3.17 Electric field in the frequency domain when (**a**) X_i is the reactant and (**b**) X_i is the product

frequency. When the properties of the unbounded media change, the continuity of the electric field requires that k_0 is conserved across a temporal discontinuity medium. Eq. (3.34) can be written as Eq. (3.50) when k_0 is conserved.

$$\mu_0\varepsilon_0\frac{\partial^2\vec{E}}{\partial t^2}+\mu_0\frac{\partial^2\phi_{X_i}(t)*\left[C_{X_i}(t)\vec{E}(t)\right]}{\partial t^2}+k_0^2\vec{E}=0 \tag{3.50}$$

When the Fourier transformation is carried out on Eq. (3.50) and $\phi_{X_i}(\omega)=\varepsilon_0\chi_{X_i}/(1+i\omega\tau_{X_i})$, the following equation is obtained:

$$\frac{k_0^2}{\mu_0\varepsilon_0}\vec{E}(\omega)-\omega^2\vec{E}(\omega)-\omega^2\frac{\chi_{X_i}(\omega)}{1+i\omega\tau_{X_i}}\frac{1}{2\pi}C_{X_i}(\omega)*\vec{E}(\omega)=0 \tag{3.51}$$

Subsequently, when the inverse Fourier transformation is carried out on Eq. (3.51), Eq. (6.21) is obtained.

$$\tau_{X_i}\frac{\partial^3\vec{E}(t)}{\partial t^3}+\frac{\partial^2\vec{E}(t)}{\partial t^2}+\tau_{X_i}\frac{k_0^2}{\mu_0\varepsilon_0}\frac{\partial\vec{E}(t)}{\partial t}+\frac{k_0^2}{\mu_0\varepsilon_0}\vec{E}(t)+\chi_{X_i}\frac{\partial^2}{\partial t^2}\left[C_{X_i}(t)\vec{E}(t)\right]$$
$$=0 \tag{3.52}$$

As k_0 is conserved, Eq. (3.52) can be rewritten as Eq. (3.53).

$$k_0^2=-\mu_0\varepsilon_0\frac{\chi_{X_i}\frac{\partial^2}{\partial t^2}\left[C_{X_i}(t)\vec{E}(t)\right]+\frac{\partial^2\vec{E}(t)}{\partial t^2}+\tau_{X_i}\frac{\partial^3\vec{E}(t)}{\partial t^3}}{\tau_{X_i}\frac{\partial\vec{E}(t)}{\partial t}+\vec{E}(t)} \tag{3.53}$$

When the parameters of the reactions change slowly compared to the microwaves, $\frac{\partial^2 C_{X_i}(t)}{\partial t^2}\vec{E}(t)$ and $\frac{\partial C_{X_i}(t)}{\partial t}\frac{\partial\vec{E}(t)}{\partial t}$ can be neglected. Eq. (3.53) can be written as follows:

$$-\frac{k_0^2}{\mu_0\varepsilon_0}=\frac{\chi_{X_i}C_{X_i}(t)\frac{\partial^2\vec{E}(t)}{\partial t^2}+\frac{\partial^2\vec{E}(t)}{\partial t^2}+\tau_{X_i}\frac{\partial^3\vec{E}(t)}{\partial t^3}}{\tau_{X_i}\frac{\partial\vec{E}(t)}{\partial t}+\vec{E}(t)} \tag{3.54}$$

As the frequency remains unchanged in this circumstance, the electromagnetic plane harmonic wave $\vec{E}(z,t)=\widehat{e}_x\exp\left[i(\omega_0 t-k_0 z)\right]$ is substituted into Eq. (3.54), which results in Eq. (3.55).

$$\frac{k_0^2}{\mu_0\varepsilon_0}=\frac{\omega_0^2\left[\chi_{X_i}C_{X_i}(t)+1+i\omega_0\tau_{X_i}\right]}{1+i\omega_0\tau_{X_i}} \tag{3.55}$$

The relation of $k_0^2=\omega_0^2\mu_0\varepsilon_0\varepsilon_r$ is substituted into Eq. (3.55), and Eq. (3.56) is obtained.

$$\varepsilon_r = \varepsilon_0 + \frac{\chi_{X_i} C_{X_i}(t)}{1 + i\omega_0 \tau_{X_i}} \tag{3.56}$$

Eq. (3.56) indicates that the dielectric behavior of the reactions can be represented by complex permittivity, which is a function of the component concentration, if the reactions satisfy the condition that the variation rate of the component concentration is slower than the variation rate of the microwaves. The results also indicate that the effective permittivity function, proposed experimentally, is reasonable under this condition (Table 3.2).

3.4.3 Multiphysics Simulation of Microwave Heating of a Slow Reaction

3.4.3.1 Modeling of the Iodination of Acetone

In this section, the iodation of acetone is carried out because it is a carefully studied chemical reaction and because significant changes in effective permittivity can be observed. The reaction is carried out at four constant temperatures of 30 °C, 37 °C, 44 °C, and 52 °C. By using the Huang method, the permittivity of a mixture during the acetone iodation reaction at 915 MHz can be obtained.

To avoid a very fast temperature increase and convection of the solution, less than 5 W power is used. The measurement is performed at 915 MHz. A coaxial probe is inserted in the solution and used to heat the reaction system. A power meter with two channels is used to measure the input and reflected power. A UMI-8 optical fiber thermometer with a diameter of 1 mm is employed to measure the temperature near the top of the coaxial probe. A PC is used to record the rising temperature. Note that the height and diameter of the beaker must be larger than two wavelengths of the microwaves in the solution in order to avoid the influence of microwave reflection off the walls of the beaker and the air–solution interface. Generally, the reflection decays rapidly due to the lossy solution. The experimental system is shown in Fig. 3.18a. According to the experiment conditions, a numerical model is used to calculate the rising temperature in the acetone iodation system, as shown in Fig. 3.18b.

The calculation result obtained by the Huang method and the experimental measurement result are shown in Fig. 3.19, where a good agreement can be observed between the measured and calculated results. This verifies the feasibility of the method. This method offers a high potential for applications in the explanation of some specific phenomena in the microwave heating of chemical reaction systems and helps in the design of future microwave reactors.

The dielectric characterization in chemical reactions is obtained by the Huang method, which can be directly used for multiphysics calculations of microwave heating.

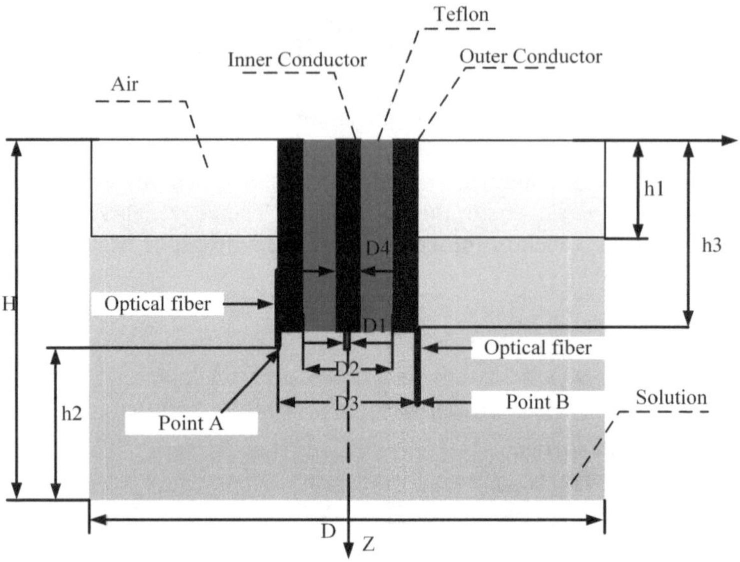

Fig. 3.18 (**a**) Experimental system. (**b**) Numerical model to simulate the microwave heating of the acetone iodation system

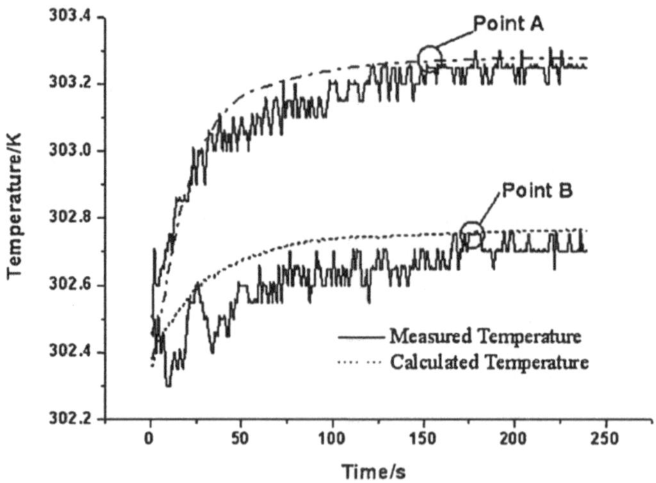

Fig. 3.19 Temperature increasing at points A and B

3.4.3.2 Modeling of Continuous Flow Processing of Biodiesel Production in a Single-Mode Microwave Cavity

In industrial applications, continuous flow systems are commonly used. Biodiesel is usually produced during a reaction between a fatty acid and alcohol. As a typical

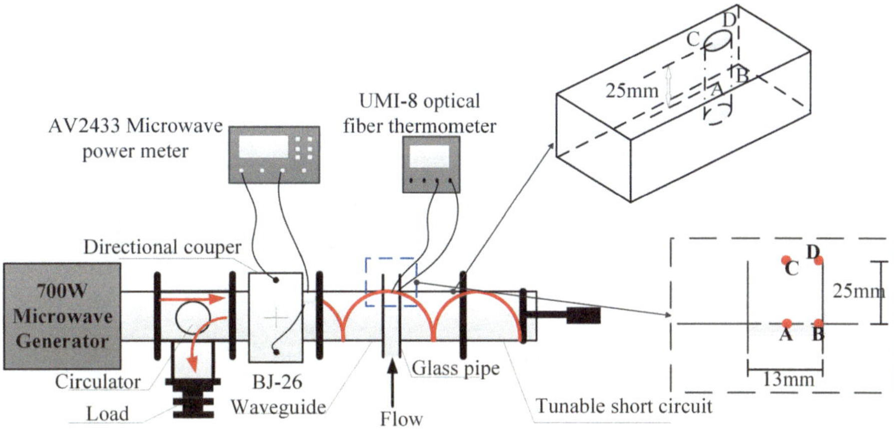

Fig. 3.20 Experiment system for the microwave-assisted continuous flow process of biodiesel production

fatty acid, in the present study, oleic acid is employed to produce biodiesel using methanol and concentrated sulfuric acid as the catalyst. The reaction equation for biodiesel production is given by

$$RCOOH + CH_3OH \underset{k_{-1}}{\overset{k_1}{\rightleftarrows}} RCOOCH_3 + H_2O \tag{3.57}$$

To enhance the yield and rate of the reaction, a volume ratio of methanol to acid of 1:1 is employed, and the mass fraction of catalyst with respect to the whole reaction solution is 1% [3]. The reagents for the synthesis reactions are obtained from local suppliers (Chengdu Kelong Chemical Reagent Factory and Chengdu Changlian Chemical Reagent Co., Ltd.). All chemicals are of analytical reagent grade.

A schematic of the experimental system is shown in Fig. 3.20. In this experiment, a glass tube (13 mm inner diameter and 17 mm outer diameter) is inserted through a BJ-26 waveguide with dimensions of 86.4 mm × 43.2 mm, along the center line, from top to the bottom. Two optical fibers are employed to measure the temperature at two locations inside the pipe (A and B), as shown in Fig. 3.20. Sites A and C are in the center of the tube, while sites B and D are near the tube's wall. The two optical fibers are connected to a FISO UMI-8 optical fiber thermometer to record a stable temperature value. Only a single value of temperature can be measured due to a very long response time of the optical fiber (~6 s for 99%) compared to the time taken by the system to reach the equilibrium temperature value for the lowest velocity of the reaction mixture (~1 s). A tunable short circuit is placed at a location in order to obtain the highest electromagnetic energy density of the standing wave in the tube inside the waveguide. The well-mixed chemical reaction solution flows into the tub from the bottom, and four different flow rates are employed.

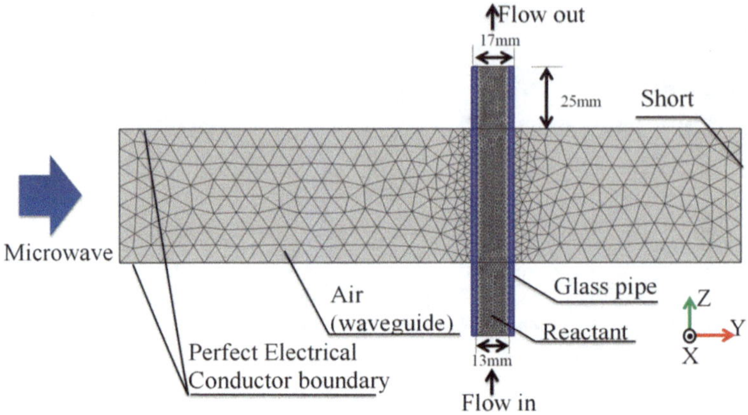

Fig. 3.21 Geometry of the microwave-assisted continuous flow process for biodiesel production

A 3D model of the above experimental system is built, and a 2D cross-section of the geometry is shown in Fig. 3.21. A BJ-26 waveguide is used as shown in Fig. 3.21. A glass pipe is employed to carry the fluid. Four out of the six faces of the waveguide are set as perfect electric conductors, except on the left and right sides faces, as shown in the figure. The left side of the waveguide is connected to a tunable short circuit to generate a standing wave in the cavity. The microwave energy comes into the waveguide from the right, and the mode of the electric field is TE_{10}. The glass tube is assumed to be a no-loss material, i.e., its dielectric loss factor is zero, but it has a complex permittivity value of 4.2. The heat transfer between the glass tube and the reaction solution is also considered to be part of the modeling framework. The reaction solution's diameter is the same as the glass pipe's inner diameter of 13 mm and the thickness of the glass pipe is 2 mm.

For electromagnetic waves, as shown in Fig. 3.21, the left side of the waveguide is connected to a tunable short circuit to generate a standing wave in the cavity. The microwave energy comes into the waveguide from the right and the mode of the electric field is TE_{10}. The other four faces are set as PEC.

For heat transfer, the bottom temperature of the initial inlet fluid into the tube is set at 303.25 K, which is the same as the room temperature. The boiling point is set at 340.20 K in the model, as the measured boiling point for this solution is 340.20 K.

During the chemical reaction, the initial inlet fluid is the well-mixed chemical reaction solution. A volume ratio of methanol to acid of 1:1 is employed, and the mass fraction of the catalyst with respect to the whole reaction solution is 1%.

As for the microwave-assisted biodiesel production system, the temperature plays a very important part in microwave heating. Figure 3.22 shows the temperature profiles at the Y-Z cross-section ($X = 54.6$) at different times (0–3 s) at four velocities. It is clear that the temperature of the reaction solution at the center of the pipe remains constant after 0.9 s. However, an increase in the solution temperature can be seen near the outlet until 1.2 s, after which the temperature remains constant until $t = 3$ s. Figure 3.23 shows the plot of the temperature profile at the

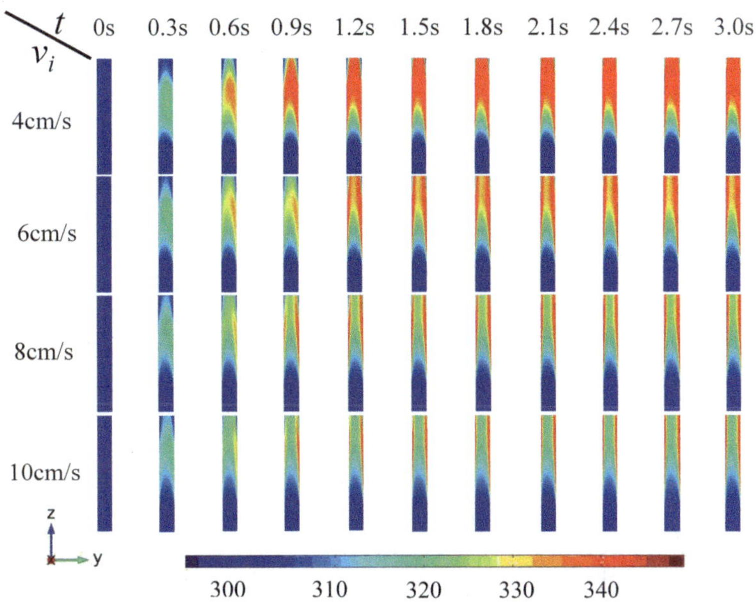

Fig. 3.22 Y-Z cross-section ($X = 54.6$ mm) temperature profile at different times and initial velocities (unit: K)

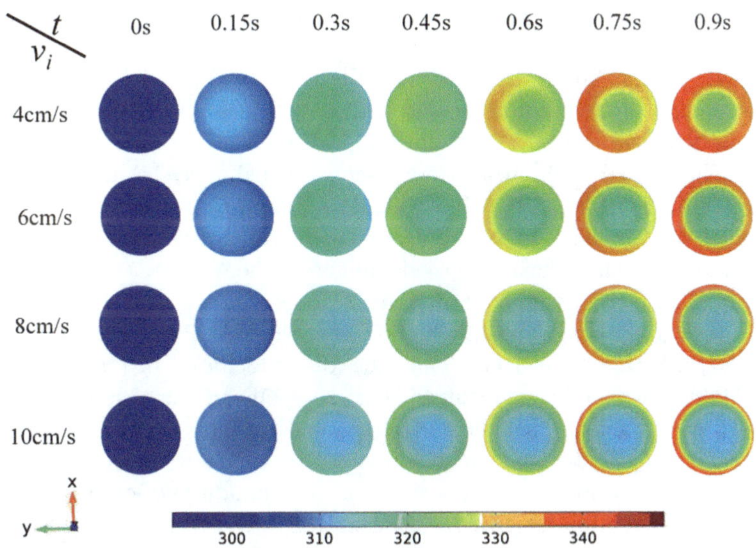

Fig. 3.23 X-Y cross-section ($Z = 27.3$ mm) temperature profile under different times and initial velocities (unit: K)

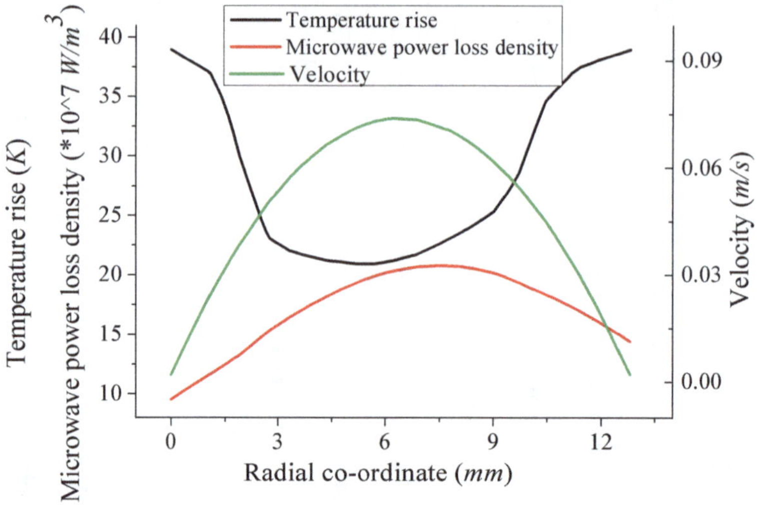

Fig. 3.24 Axial velocity profile, temperature rise profile, and electromagnetic power loss density profile (cross line from site [54.6, 143.5, 27.3] to site [54.6, 156.5, 27.3]) at 0.9 s under the 4 cm/s initial velocity

X-Y cross-section (Z = 27.3 mm) at the center of the tube from 0 to 0.9 s. After 0.9 s, the temperature remains constant, as indicated in Fig. 3.23. The figures indicate that the temperature near the walls is higher than that at the center. To explain this phenomenon, the axial velocity profile, temperature rise, and electromagnetic power loss density profile (line from coordinate site [54.6, 143.5, 27.3] to [54.6, 156.5, 27.3]) at 0.9 s at 4 cm/s initial velocity are plotted in Fig. 3.24. Even though the power loss is higher near the center, the temperature rise is not as much as it is near the wall. This is due to the heat energy being carried away because of the much higher velocities near the center. However, even though the power loss near the walls is relatively smaller than that at the center, the solution is heated to much higher temperatures due to the much lower velocities near the walls.

The simulation results indicate that the average stable temperature of site A is 340.86 K (flow rate: 4 cm/s), 330.96 K (flow rate: 6 cm/s), 324.14 K (flow rate: 8 cm/s), and 320.06 K (flow rate: 10 cm/s) for the four velocities. The maximum relative error is 0.908%. The predicted results match very well with the experimental results. This validation indicates that the model can be used to describe the process of microwave heating of the biodiesel reaction.

Here, the nonuniform water concentration distribution and temperature can be observed for the continuous process, especially at lower flow velocities. From the simulation results, the water concentration and temperature are relatively higher near the wall than at the center. This is largely due to the very low velocity near the wall, and this effect will have a strong influence on large-scale processing. Therefore, the improvement in uniformity should be considered. In this study, by considering the heating uniformity and energy saving, the best flow rate is 320 ml/min. The uniformity optimization obtained by multiphysics simulation for large-scale processing can be based on this work. Since velocity plays a major role in this process, optimization should start with solving the nonuniformity of the velocity field.

3.4.3.3 Dynamic Analysis of Continuous-Flow Microwave Reactor Based on a Screw Propeller

Continuous-flow processing is a popular approach to overcome the apparent limitations of large-scale microwave processing. However, inhomogeneous heating still exists in the microwave continuous applicator and stirring becomes a common and effective method to overcome this issue. Furthermore, a detailed thermal analysis of the microwave heating process is important to control the reaction conditions and confirm the reaction progress. However, the temperature inside the applicator is difficult to measure under a stirring condition. This necessitates the use of a multiphysical field simulation of the heating process in order to understand the heating dynamics. Meanwhile, a screw propeller is a commonly used device in aircraft and material handling systems. When applied in a continuous flow reactor, its rotation will not only stir the material but also change the electric field distribution, which has a great potential in improving the microwave heating uniformity. Therefore, employing a screw propeller in the microwave continuous-flow applicator and simulating the heating process will be meaningful. Unfortunately, due to the computational difficulty of a microwave heating process with moving elements and the complexity of the movement of the fluid under stirring, the numerical simulation of the microwave heating process that considers stirring has not been reported. In this study, a step-by-step algorithm based on an implicit function, level set methods, and ALE machinery is developed to compute the microwave heating process with a screw propeller, which will be helpful for the analysis and design of microwave reactors that employ stirring.

A schematic of the experimental system is shown in Fig. 3.25. The effective microwave power (1255 W) is measured using a directional coupler and microwave power meter. The flow rate of water (27.8 mL/s) is controlled using a peristaltic pump. At the outlet, the flow is divided into eight tributaries, and each tributary flows to the corresponding position in the fluid collection platform according to its number, as shown in Fig. 3.25. To avoid the effects of gravity, the fluid collection platform is placed at the same height as that of the fluid inlet. During the experiments, we first fill the Teflon pipe with water by using a peristaltic pump, and then open the microwave generator, peristaltic pump, and motor simultaneously. Finally, the equipment is turned off after heating for 5 min. The temperature of the eight tributaries is measured by the corresponding number of optical fibers and is recorded at an interval of 10 s by using an optical fiber thermometer.

A geometric model for the multiphysical field computation is developed based on the experimental system. The heating process is computed by the commercially available multi-physics software. Liquid materials used in this communication are in-compressible Newtonian fluids, and the maximum Reynolds number (calculated

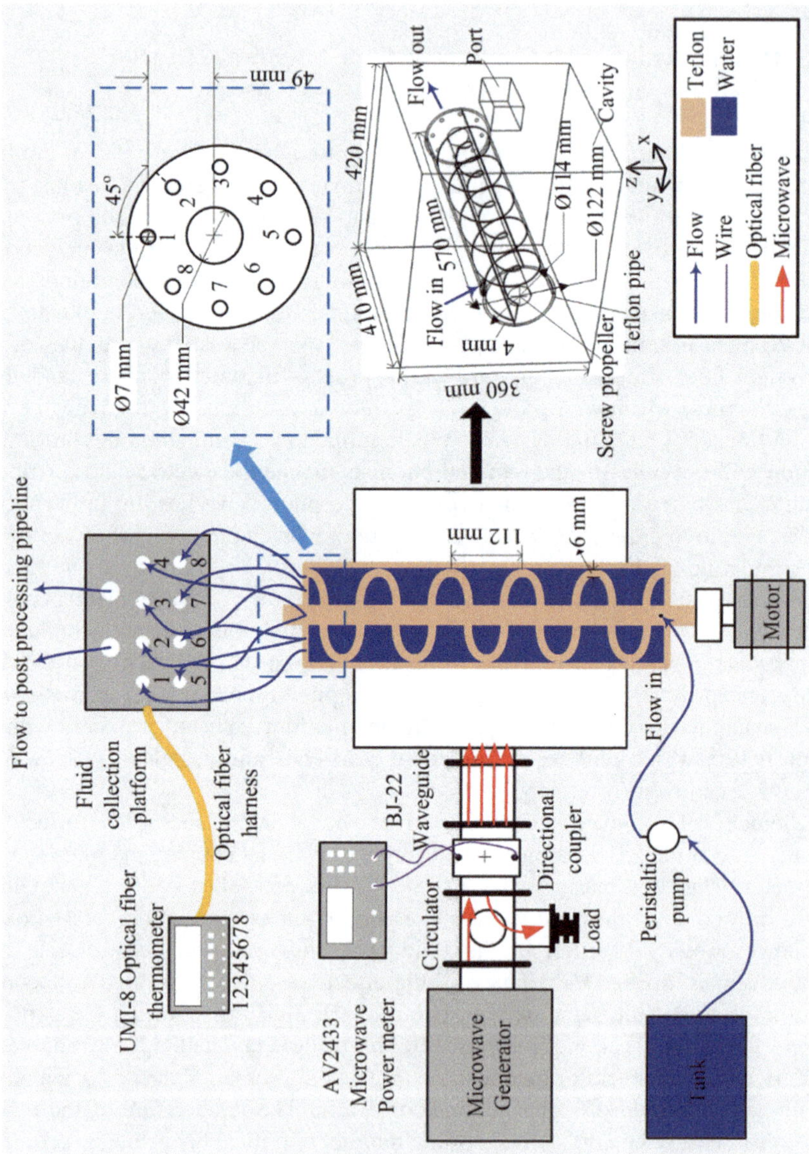

Fig. 3.25 Schematic of the experimental system

by Eq. (3.58)) of the liquids is small (1083). Ren is smaller than 1400, therefore, the fluid flow inside the pipe is treated as laminar.

$$Ren = \rho N D^2 / \mu \tag{3.58}$$

where the screw propeller speed, N, is 5 rpm; the screw propeller diameter, D, is 0.114 m; and ρ and μ are the density and viscosity of the liquid, respectively. The liquid sticks to the walls of the screw and barrel, and thus, requires a "no-slip" boundary condition at the walls. Table 3.3 summarizes the material properties used for the computations. The initial temperature is 15 °C.

A step-by-step algorithm is adopted to simulate the heating process, as described in the calculation flowchart in Fig. 3.26. In each step, the calculation of the electromagnetic field is separated from the calculation of the temperature and velocity fields. The electromagnetic field is calculated in the frequency domain and governed by Maxwell's wave form.

The velocity and pressure distribution of the fluid are governed by the Navier-Stokes equation and the mass conservation equation (continuity equation) (3.62).

$$\rho \cdot \frac{\partial u}{\partial t} + \rho(u \cdot \nabla)u = \nabla \cdot \left\{ -p\vec{I} + \mu\left[\nabla u + (\nabla u)^T\right] - \frac{2}{3} \cdot \mu(\nabla \cdot u)\vec{I} \right\} \tag{3.59}$$

$$\frac{\partial \rho}{\partial t} + \nabla \cdot \left(\rho \vec{\mathrm{u}}\right) = 0 \tag{3.60}$$

where p is the pressure and $\vec{I}$ is the unit matrix. The stirring process caused by the screw propeller is achieved by invoking the Arbitrary Lagrangian-Eulerian Formulation (ALE) machinery.

$$d\vec{x} = d\vec{x}\left(r_{bp}, \omega, t\right) \frac{d\omega}{dt} = w \tag{3.61}$$

where $\vec{\mathrm{r}}_{bp}$ is the rotation axis vector, $\vec{x}$ is the position vector, and ω is the angular displacement, w is the angular velocity.

The temperature and velocity fields are coupled and calculated in the time domain. Although the rotation of the screw propeller, in the range of 30°, has little effect on the electric field, its rotation alters the shape of the liquid, thereby changing the heat source distribution. To incorporate this phenomenon, an implicit function is used to represent the position of the water and screw propeller. Using the screw propeller shaft as the x-axis, the initial position of the screw propeller is shown as follows:

$$\begin{aligned} \Phi_s(x, y, z, 0) = {} & \left(\sqrt{y^2 + z^2} < r_o\right) \cdot \left(\sqrt{y^2 + z^2} > r_i\right) \cdot [y/z > \tan(2\pi x \eta / l)] \cdot \\ & [y/z > \tan(2\pi(x + h)\eta / l)] + \left(\sqrt{y^2 + z^2} < r_i\right) \end{aligned} \tag{3.62}$$

Table 3.3 Summary of the material properties applied in the simulation

Parameter	water	Ethylene glycol	Glycerol	Teflon
Dielectric constant (ε')	80	37 6.	33	2.1
Loss tangent ($\tan\delta$)	0.18	1.35	0.54	0
Specific heat capacity (C_P, $J/(kg \cdot K)$)	4187	2383	2385	1050
Density (ρ, kg/m^3)	1000	1116	1246	2200
Thermal conductivity (k, $W/(m^3 \cdot K)$)	0.59	0.25	0.29	0.24
Dynamic viscosity (μ, $Pa \cdot s$)	$1.38–0.02\ T$ $+1.36E^{-4}T^2 - 4.65E^{-7}T^3$ $+ 3.27E^{-3}T^2 - 6.63E^{-6}T^3$ $+8.90E^{-10}T^4 - 9.08E^{-13}T^5$ $+3.85E^{-16}T^6$	$58.77–0.72\ T$ $+3.27E^{-3}T^2 - 6.63E^{-6}T^3+5.05E^{-9}T^4$	$318121.21–5155.23\ T$ $+33.437T^2 - 0.11T^3 + 1.76E^{-4}T^4$ $- 1.14E^{-7}T^5$	

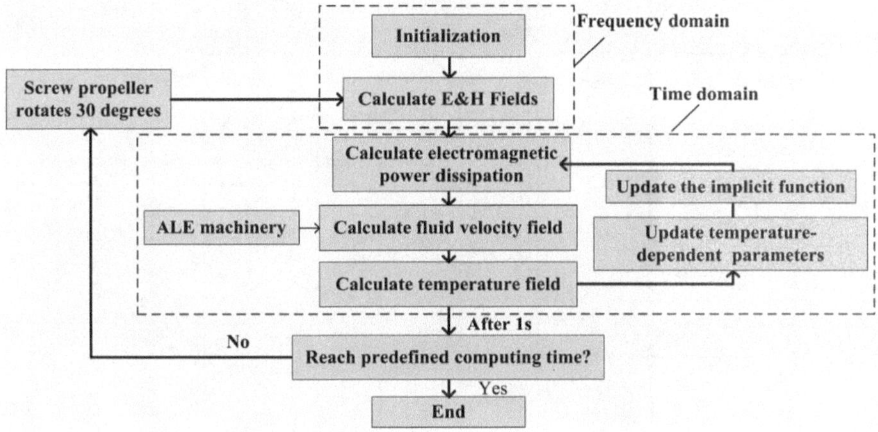

Fig. 3.26 Calculation flowchart

where r_o, r_i, η, and l are the inner radius, outer radius, turn number, and length in the x-direction of the screw propeller, respectively. h is the thickness of the screw blade. When the screw propeller is rotating, by using the level set method, the position can be expressed as

$$\Phi_s(x,y,z,t) = \left(\sqrt{y^2+z^2} < r_o\right) \cdot \left(\sqrt{y^2+z^2} > r_i\right) \cdot [y/z > \tan((2\pi x\eta + w \cdot t)/l)] \cdot [y/z > \tan((2\pi(x+h)\eta + w \cdot t)/l)] + \left(\sqrt{y^2+z^2} < r_i\right) \tag{3.63}$$

where ω is the angular velocity of the screw propeller. Based on this, the loss tangent of the liquid during heating can be expressed as

$$\tan\delta_h(x,y,z,t,T) = \left[\left(\sqrt{y^2+z^2} < r_p\right) - \Phi_s(x,y,z,t)\right] \cdot \tan\delta_1(T) \tag{3.64}$$

where r_p is the inner diameter of the Teflon pipe and $\tan\delta_1(T)$ is the loss tangent of the liquid. Then, by using the obtained electric field, the local power dissipated inside the pipe can be obtained from Eq. (3.3).

The temperature distribution inside the pipe is governed by heat conduction in the solid and heat conduction and convection in the fluid.

The temperature of the eight tributaries obtained by simulation and that obtained experimentally are presented in Fig. 3.27, which clearly indicates that the outlet temperature of the liquid reaches a steady state after heating for about 250 s. However, the simulation results reach a steady state earlier than the experimental results, which may be due to the instability of the power source. Nevertheless, the measured temperature is in good agreement with the simulation results, which

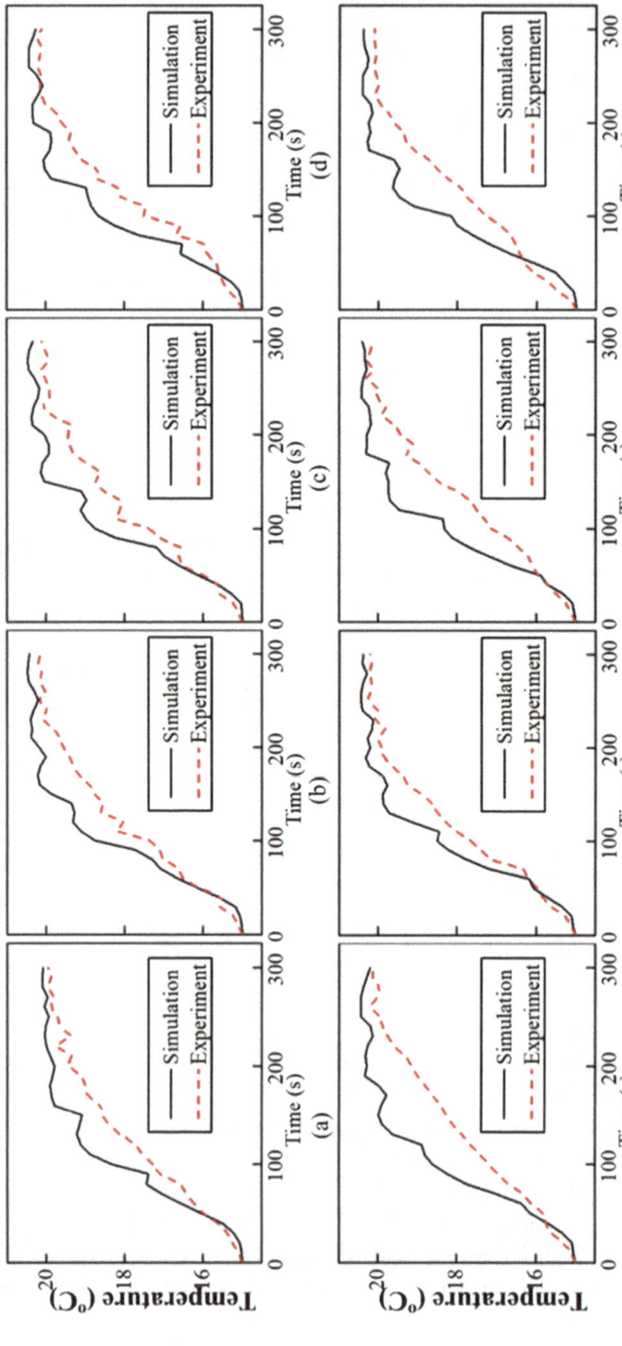

Fig. 3.27 Temperature of the eight tributaries obtained by simulation and experiments, (**a**)–(**h**) correspond to the 1–8 tributaries in turn

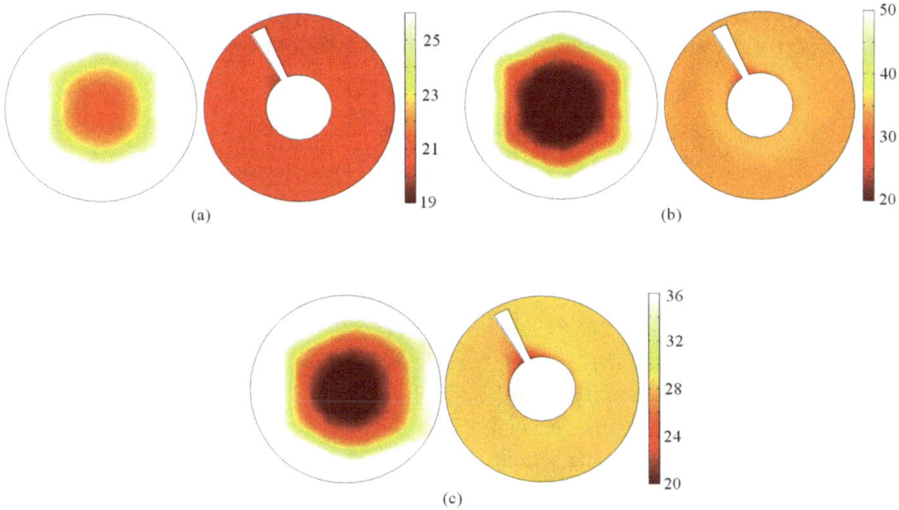

Fig. 3.28 Outlet temperature distribution at a steady state with and without a screw propeller (Unit: °C). (**a**) Water. (**b**) Ethylene glycol. (**c**) Glycerol

verifies that the model can be applied to the calculation of microwave heating with stirring in a continuous-flow reactor.

To further analyze the effect of the screw propeller on the heating uniformity of other commonly used liquids, the heating process of ethylene glycol (with a viscosity about 20 times that of water) and glycerol (with a viscosity about 1000 times that of water) are simulated. The outlet temperature distributions when reaching the steady state under the same conditions with or without a screw propeller are compared in Fig. 3.28. This shows that the uneven temperature distribution between the inner and outer areas of the liquids inside the pipe can be improved well by using a screw propeller. When the temperature nonuniformity is quantified by the coefficient of variation (the ratio of standard deviation to mean), the improvement in the temperature uniformity of water, ethylene glycol, and glycerol contributed by the screw propeller is 88%, 96%, and 94%, respectively.

3.4.4 Multiphysics Simulation of Microwave Plasma

3.4.4.1 Numerical Investigation of the Surface Wave Formation in a Microwave Plasma Torch

Plasma modeling is an exciting subject in which virtually all physical disciplines are represented. The plasma model combines the electromagnetic, statistical, and fluid dynamic theories. Generally, plasma can be considered as either a dielectric with the relative permittivity ε_{rp}, or equivalently, a conductor with conductivity σ_p. μ_r is the relative permeability and is equal to 1 for a plasma without a static magnetic field:

$$\sigma_p = \frac{n_e q^2}{m_e(\nu_m + j\omega)} = \frac{\varepsilon_0 \omega_p^2}{\nu_m + j\omega} \tag{3.65}$$

$$\varepsilon_{rp} = 1 - j\frac{\sigma_p}{\omega\varepsilon_0} = 1 - \frac{\omega_p^2}{\omega^2 + \nu_m^2} - j\frac{\omega_p^2 \nu_m}{\omega(\omega^2 + \nu_m^2)} \tag{3.66}$$

$\omega_p = \left(\frac{n_e q^2}{m_e \varepsilon_0}\right)^{\frac{1}{2}}$ is the electron plasma frequency, which can be approximated as the plasma oscillation frequency. n_e is the electron density. ν_m is the electron elastic collision frequency with neutral atoms and ions. q and m_e are the elementary charge and mass of a single electron, respectively. ε_0 is the vacuum permittivity constant. Having solved the Helmholtz equation in the frequency domain, the plasma-absorbed microwave power density can be expressed as

$$Q_{E_z=1} = \frac{1}{2}\,\mathrm{Re}\left(\sigma\vec{E}\cdot\vec{E}^*\right) \tag{3.67}$$

Here, Re denotes the real part of the expression and $*$ denotes the complex conjugate. Note that the total plasma-absorbed microwave power, P_{total}, deposited into the plasma column is a given parameter

$$K_Q = \frac{P_{total}}{\int_V Q_{E_z=1}\,dV} \tag{3.68}$$

With the normalization parameter of K_Q, the total spatial microwave loss power density, $Q_{EM,}$ is transferred to the electrons and can be written as follows:

$$Q_{EM} = K_Q \times Q_{E_z=1} \tag{3.69}$$

Typically, there are three main types of mathematical models to describe a plasma: fluid approximation [10], kinetic approach [11], and hybrid method [12]. Due to its high efficiency and large-scale calculation capability, fluid approximation is chosen in our case [13]. The number densities of ions and excited atoms are obtained by solving the following continuity equation:

$$\nabla\cdot(-D_k\nabla n_k) + (\vec{\mu}\cdot\nabla)n_k = R_k \tag{3.70}$$

where n_k (unit: $1/\text{m}^3$) is the number density of each species in the plasma mixture and its index k indicates the excited neutral atoms Ar (4 s) (k = s) and Ar (4p) (k = p), atomic ions Ar^+ ($k = i_1$), and molecular ions $Ar_2{}^+$ ($k = i_2$). Note that the argon neutral atom Ar (N_n) in the ground state is obtained via the ideal gas law. R_k is the species source term dominated by the plasma chemistry. $\vec{\mu}$ is the plasma bulk flow velocity that contributes to the forced species convection and to the species loss. D_k is the diffusion coefficient for species k. Here, the effective diffusion coefficients are taken for excited atoms [14] and the ambipolar diffusion coefficients are applied to ions [15] (responsible for space charge and quasi-neutrality).

The electron density is approximated as the summation of the positive charge densities under the assumption of quasi-neutrality:

$$n_e = n_{i1} + n_{i2} \tag{3.71}$$

Correspondingly, the electron diffusion flux can be described by

$$\vec{\Gamma_e} = -D_{i1}\nabla n_{i1} - D_{i2}\nabla n_{i2} \tag{3.72}$$

A two-temperature model is applied to describe the energy conservation in the system. The temperatures of the heavy species (T (unit: K)) and the electrons (T_e(unit: K)) are modeled independently ($T \neq T_e$):

$$\nabla \cdot \overrightarrow{\Gamma_{en}} + (\vec{\mu} \cdot \nabla)\left(\frac{3}{2}T_e K_B n_e\right) = Q_{EM} - Q_{el} - Q_{inel} \tag{3.73}$$

Eq. (3.73) describes the energy conservation of the electrons, where the second term on the left-hand side describes the forced convection introduced by the bulk flow. $\overrightarrow{\Gamma_{en}}$ is the electron energy flux determined by the electron flux, and the electron energy thermal diffusion and is given by

$$\overrightarrow{\Gamma_{en}} = \frac{5}{2}K_B T_e \vec{\Gamma_e} - K_e \nabla T_e \tag{3.74}$$

where K_e is the electron thermal conductivity. Q_{EM} is the power density determined from the electromagnetic field according to Eq. (3.69). Q_{el} is the power density transferred from the electrons to the neutral atoms and the ions via elastic collisions, and is given by

$$Q_{el} = \frac{3}{2}n_e K_B \frac{2m_e}{m_{A_r}} \nu_m (T_e - T) \tag{3.75}$$

Here, K_B is the Boltzmann's constant and m_{A_r} is the mass of an argon atom. T is the temperature of the heavy species.

The calculations start with the assumed initial values for different species' continuity equations, together with a given gas inflow rate and a total microwave deposited power. The procedure for the entire numerical implementation is presented in Fig. 3.29.

The total number of the mesh elements is 11,827. This mesh density is validated by comparing the results obtained from a much denser mesh structure. The model is

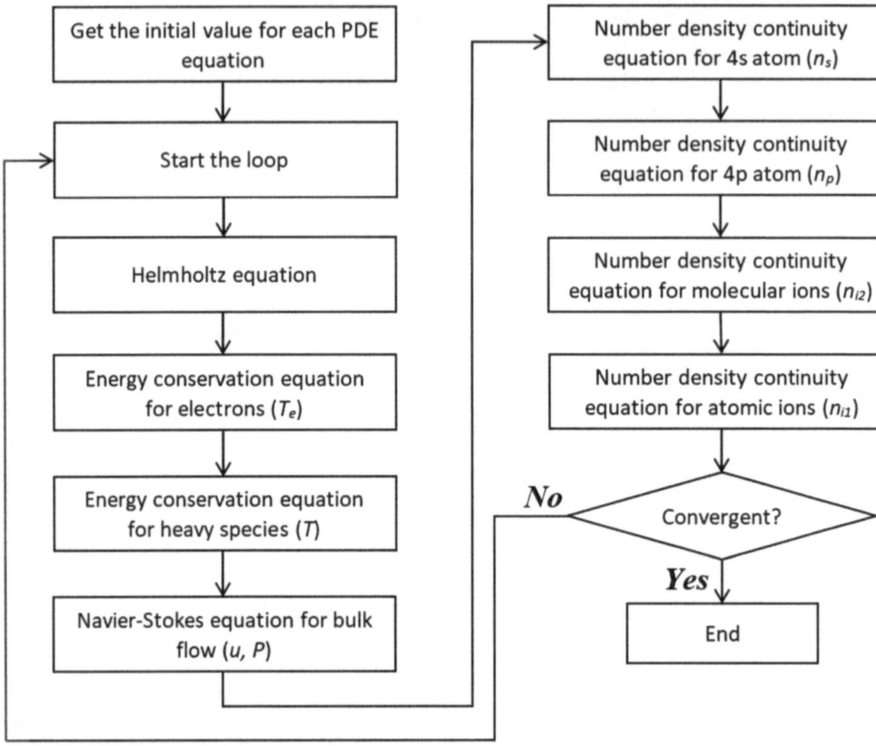

Fig. 3.29 Calculation procedure

solved on a computer with an i7-5500U CPU and 16 GB RAM. It takes approximately 45 min to complete the entire calculation for one model.

As shown in Fig. 3.30, there is an increase in the plasma-absorbed microwave power.

The electron number density changes in the following three aspects: (1) the electron number density increases significantly; (2) the higher electron number density region in the plasma column gradually moves closer to the glass tube walls; and (3) the higher electron density region in the plasma column becomes increasingly longer in the axial direction. The extension of the higher electron density region in the axial direction implies that the plasma column is confined within the interaction area between the cylindrical glass tube and the rectangular waveguide, when the total plasma-absorbed microwave power is low (200 W). However, as the total plasma-absorbed microwave power increases, the plasma column stretches and extends out from the rectangular waveguide area.

These changes in the electron number density, value and profile, will finally result in variation of the electromagnetic properties of the plasma column. Re-examining the plasma relative permittivity expression (3.66), we can see that the real part of ε_{rp} becomes negative when $\omega_p^2 < \omega^2 + v_m^2$. If the plasma is collisionless ($v_m = 0$), the

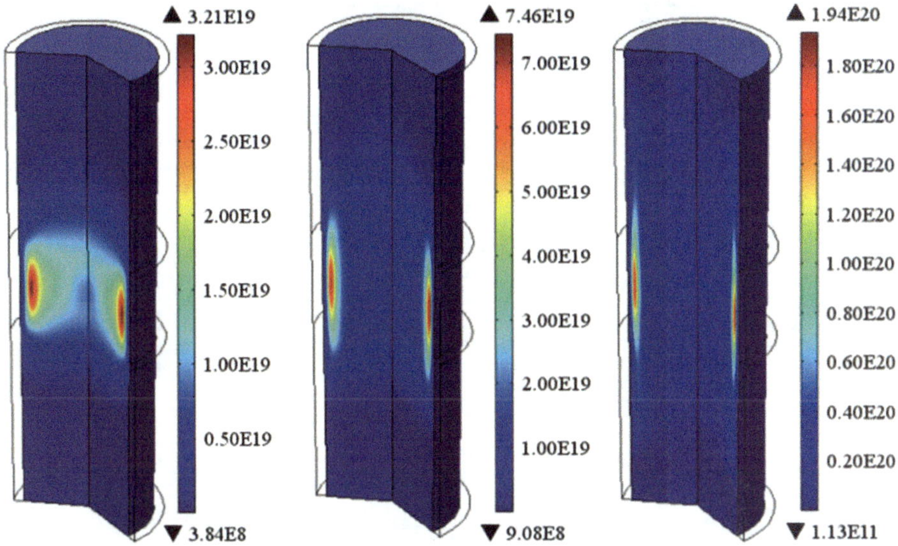

Fig. 3.30 Electron number density (units: $1/m^3$) for different plasma-absorbed microwave powers ($\varphi = 1$ L/min). (**a**) 200 W. (**b**) 400 W. (**c**) 800 W

electron density, n_e, has a critical number density of approximately 7.6×10^{16} [$1/m^3$], which corresponds to $\omega_p = \omega$ and real $(\varepsilon_{rp}) = 0$. In our case, the plasma is highly collisional. Thus, the critical electron number density should be even higher than 7.6×10^{16} [$1/m^3$]. If the plasma-absorbed microwave power continues to increase, the electron number density in the plasma column will eventually exceed this threshold, thereby turning the real part of the plasma relative permittivity into a negative value. In this case, the plasma relative permittivity will have a negative real part and a large imaginary part. The plasma column will behave as a conductor. Microwaves can penetrate the plasma within only a small length. Figure 3.31 presents the radial distribution of the microwave electric field intensity, which can illustrate the dependence of the microwave shielding ability of the plasma column on the plasma-absorbed microwave power.

Figure 3.31 shows that the intensity of the electric field reaches a considerably higher value at the interaction boundary of the plasma column (conductor) and the glass. Subsequently, it attenuates quickly within a small depth in the plasma column. The larger the microwave power the plasma absorbs, the more conductive the plasma column is. In addition, the more harder it is for the microwave to penetrate the plasma column, the stronger the capacity of the plasma column becomes to shield the microwave.

Due to the microwave shielding, the electrons in the skin depth area obtain much more microwave energy compared to those in the other regions of the plasma column, which result in a larger ionization rate and a higher electron number density in the skin depth area. Thus, a decrease in the microwave penetration depth will lead

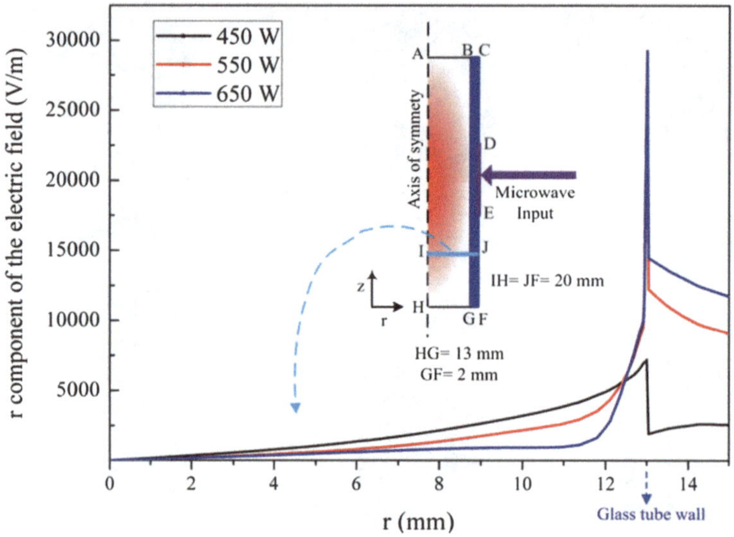

Fig. 3.31 Intensity of the microwave electric field *r* component (abs(*Er*)) for different plasma-absorbed microwave powers ($\varphi = 1$ L/min)

to a higher electron density region in the plasma column to move gradually close to the inner glass tube wall, as shown in Fig. 3.30.

All the above analyses demonstrate that the plasma column becomes increasingly conductive or reflective with an increase in the plasma-absorbed microwave power. However, being more conductive does not absolutely imply a greater reflection of the input microwave power. This is due to the formation of a "coaxial waveguide" and the traveling surface wave. Either before the discharge or when the plasma-absorbed microwave power is low enough, the electron number density can be lower than the critical electron number density. Thereby, the plasma column is not sufficiently conductive to be considered as a conductor, and the relative permittivity of the material in the discharge glass tube can be in the range between 0 and 1. In this case, the cylindrical discharge tube is a cutoff circular waveguide for the input microwave at a frequency of 2.45 GHz. The calculation expression of the cutoff frequency for the lowest electromagnetic mode, TE_{11}, of a circular waveguide is given by

$$f_{circular} = \frac{1.8412c}{2\pi R} \tag{3.76}$$

where R is the inner radius of the discharge tube and c is the speed of light in vacuum. The cut-off frequency of the cylindrical discharge tube under the current size is estimated as 5.8607 GHz, without considering the influence of the glass and 2.45 GHz is the working frequency of the MPT, which is lower than the cutoff frequency. Thus, the discharge tube is cut off from the input microwave and no propagating electromagnetic mode exists when the electron number density is lower than the critical number density.

However, when the plasma-absorbed microwave power is sufficiently high, the plasma column can be treated as a good conductor. Consequently, the plasma column and the dielectric glass tube, together with the surrounding cylindrical metallic walls, form a "coaxial waveguide." The cylindrical metallic wall is the outer conductor of the "coaxial waveguide," the plasma column acts as the inner conductor, and the glass tube is the dielectric between the inner and outer conductors.

The "coaxial waveguide," as a two-conductor transmission line, contains the transverse electromagnetic (TEM) mode. The special characteristic of the TEM mode enables the input microwave to propagate in the discharge tube without the cutoff frequency limitation. Thus, once the "coaxial waveguide" is formed, the input microwave can change its propagation direction, enter the discharge tube, and propagate along the plasma column toward both the gas inlet and the outlet. According to our analysis above, the surface wave is identified as the traveling microwave in the TEM mode in the "coaxial waveguide."

In addition, once the "coaxial waveguide" is formed, the interaction region between the discharge tube and the rectangular waveguide acts as a "mode adapter," which converts the rectangular waveguide mode (TE_{10}) to the coaxial waveguide mode (TEM). Based on the mode adapter theory, we can infer that the return loss of the MPT due to the reflection of the plasma column can be improved by optimizing the design of the "rectangular waveguide to coaxial mode adapter." More precisely, the inner and outer radii of the glass tube and its dielectric coefficient can determine how much of the input microwave power will be reflected after igniting the plasma "flame" under a constant gas inflow rate. This is why the experiment in [16] found that the thickness of the discharge tube plays an important role in the plasma column length.

3.4.4.2 Multiphysics Computation for Plasma Torch Design

1. Tapered waveguide design in plasma
   The WR-430 waveguide (109.2 mm × 54.6 mm) used in the microwave plasma torch was tapered to a reduced cross-section of 109.2 mm × 10 mm, a reduction of one wavelength in the frequency of 2.45 GHz. The simulation models are illustrated in Fig. 3.32 by the top and side views.

The generated microwave radiation is guided through the tapered waveguide, and enters the quartz discharge tube with an outer diameter of 50 mm. The tube outside the waveguide is enveloped by a metal sleeve in order to prevent energy leaks, as shown in Fig. 3.32a. The characteristic impedance of the tapered WR-430 waveguide is matched by tuning the angle of the four metal sheets (Fig. 3.32b).

It is known that the plasma is excited when the field strength reaches the threshold (E_c) of the discharged gas. E_c has the form [17, 18]

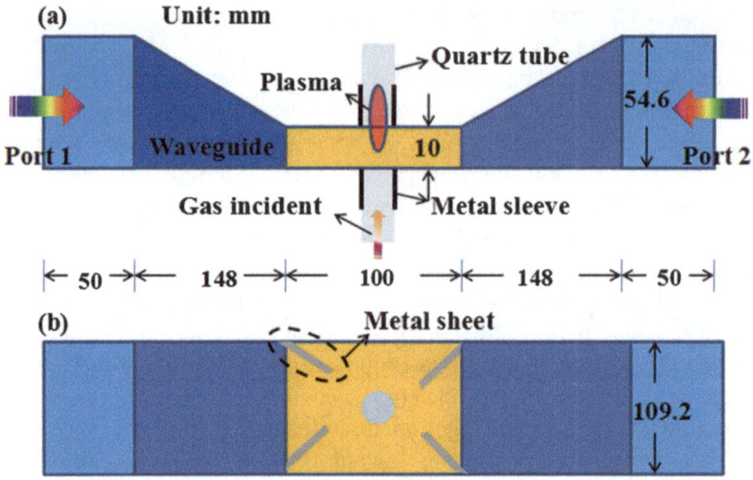

Fig. 3.32 Schematic shows the waveguide plasma apparatus. (**a**) Top view and (**b**) side view

$$E_c = 2.415 \times 10^{20} m_e(\nu_{en}/n_N)P\sqrt{2U_i/(e^2M)}Vm^{-1} \tag{3.77}$$

Here P represents the atmospheric pressure, U_i denotes the ionization energy, e is the elementary charge, m_e is the electron mass, and ν_{en}/n_N indicates the electron−neutron collision rate. Under atmospheric pressure, argon has the value of $\nu_{en}/n_N = 2.4 \times 10^{-13} m^3 s^{-1}$ and $U_i = 15.755$ eV. Then, the threshold (E_c) of the discharged argon can be calculated as 3×10^5 V/m. Argon is selected as the working gas duc to its stability and because it is easier to excite than nitrogen.

The simulated model is optimized to the best dimensions by finite-element analysis software. The microwave field of the discharge tube is observed intuitively, and is agreed well with the computed results. Simulations are divided into three cases: (1) Port 1 input has a total power of 500 W and port 2 is set as the terminated end, or conversely. (2) Ports 1 and 2 both have an input of 250 W of coherent power. (3) Ports 1 and 2 both have an input of 250 W of incoherent power. The incoherent source is represented by a random signal as

$$E(t) = E_0 U(t) \sin\left[2\pi(f \pm Ran1)t + Ran2\right] \tag{3.78}$$

Ran1 denotes the random change in the frequency, which is a random number within -50×10^6 to 50×10^6. Meanwhile, Ran2 denotes the random change in the phase, which is a random number within $0 - 2\pi$. U(t) denotes the raised cosine function. The following discussion of electric-field distribution are conducted under condition of no flow and flow field.

Without flow means that it does not refer to continuity and momentum equations, heat transfer equations, and the general species equation. Electric-field distributions under the three cases of input are shown in Fig. 3.33. Figure 3.33a displays a single

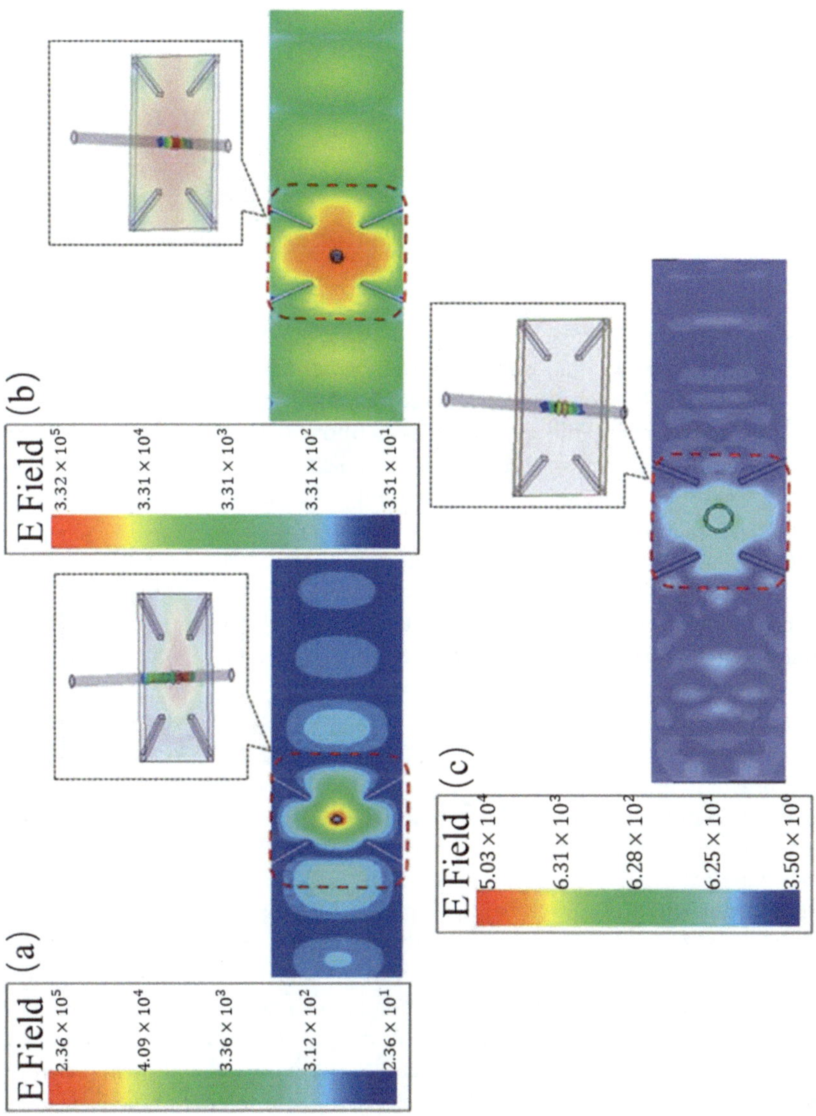

Fig. 3.33 Electric-field distributions (V/m) under three different input cases. **(a)** Single port, **(b)** dual port of coherent power, and **(c)** dual port of incoherent power

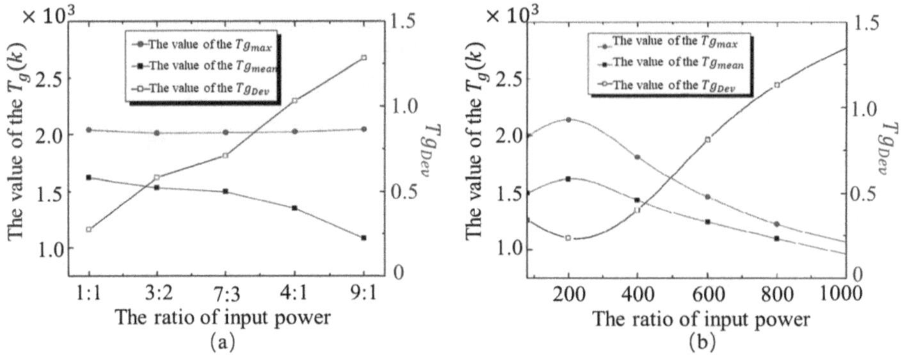

Fig. 3.34 Values of gas temperature with variation in (**a**) the ratio of input power and (**b**) gas flow rate

port of the input, whose E_{max} reaches 2.36×10^5 V/m, but it fails to exceed the Ec of argon. Figure 3.33c shows a dual-port incoherent power input, whose E_{max} is far less than E_c. By using a coherent power-combining technique, E_{max} reaches 3.32×10^5 V/m and meets the discharged condition as shown in Fig. 3.33b. In contrast to the above two cases, the $E_{\max}$ is increased by 0.5 and 5 times, respectively. Therefore, the coherent power-combining technique will maximize the electric-field strength and minimize the waste of energy.

To study the optimum conditions for an industrial application, a 2D argon fluid model based on the above coherent power-combining technique is presented. The ratio of input power and gas flow rate is varied, and the variations in the gas temperature are obtained. Figure 3.34 shows the different values of the gas temperature as the microwave power or gas flow rate is varied. T_{gmax}, T_{gmean}, and T_{gDev} denote the maximum value, mean value, and average variance of the gas temperature, respectively. In Fig. 3.34a, when the total input power is 1 kW, an increase in the ratio of the input power means that T_{gmax} remains almost constant. There appears an apparent decline in T_{gmean} and an apparent rise in T_{gDev} when the ratio is greater than 7:3, which is detrimental to the uniformity of the plasma. As shown in Fig. 3.34b, when the ratio of the input power is 1:1, the gas flow rate is greater than 200 mL/min and T_{gmax} and T_{gmean} decrease by degrees, but T_{gDev} behaves inversely. This indicates that the increase in the gas flow rate leads to a reduction in the electron transfer and an acceleration of heat transfer, which enlarges the nonuniformity of the plasma temperature. For an increase in the gas flow rate below 200 mL/min, there is an inverse trend in the gas temperature behavior. This is because the amount of gas molecules decreases with gas discharge. Once the rate becomes too small, the lack of gas supply results in the termination of the discharge, which is not beneficial for a sustained plasma. In summary, a stable and high-density plasma is obtained with a port ratio of 1:1 and a flow rate of 200 mL/min.

To prove the superiority of coherent power-combining over the conventional single-port technique in a flow field, the electric-field distribution, electron density, and gas temperature are simulated for different cases of the input. For the condition

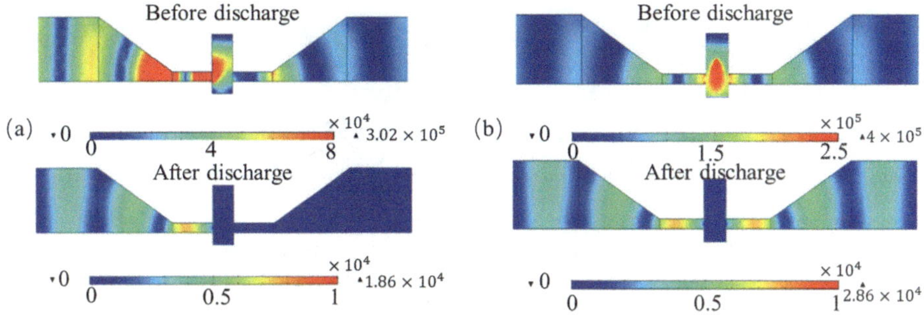

Fig. 3.35 2D distributions of electric field (V/m) under single port (**a**) and dual port (**b**), with coherent power-combining before and after discharge

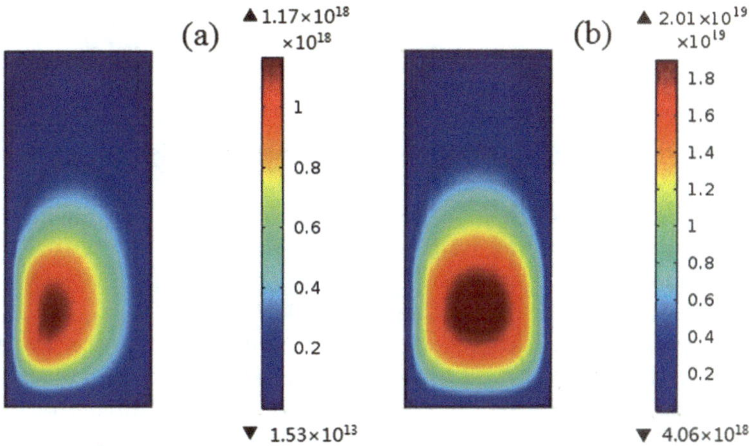

Fig. 3.36 2D distributions of electron density (m^3) under single port (**a**) and dual port (**b**), with coherent power-combining after discharge

where the total input power is 1 kW and the flow rate is 200 ml/min, the electric-field distributions before and after the discharge are shown in Fig. 3.35. Figure 3.35a and b represent the field distribution under a single port and a dual port with coherent power-combining, respectively. Before discharge, E_{max} and E_{mean} reach 4.0×10^5 V/m and 2.97×10^5 V/m, respectively, by using the coherent power-combining technique. In contrast to a single port, these values are increased by nearly 1×10^5 V/m. Furthermore, the electric field is more uniform in the discharged area, which indicates an effective absorption of microwave energy. As shown, from the time of 0.001 s, the phenomenon of gas discharge appears in the quartz tube, which shortens the time and promotes the efficiency compared to a single-port structure. After the discharge, the plasma behaves like a metal, which the microwave cannot penetrate. Figure 3.36 shows the 2D distribution of the electron density after discharge. As shown in Fig. 3.36a for a single port, some of the electrons are concentrated on the edge of the tube, which behaves nonuniformly for a plasma.

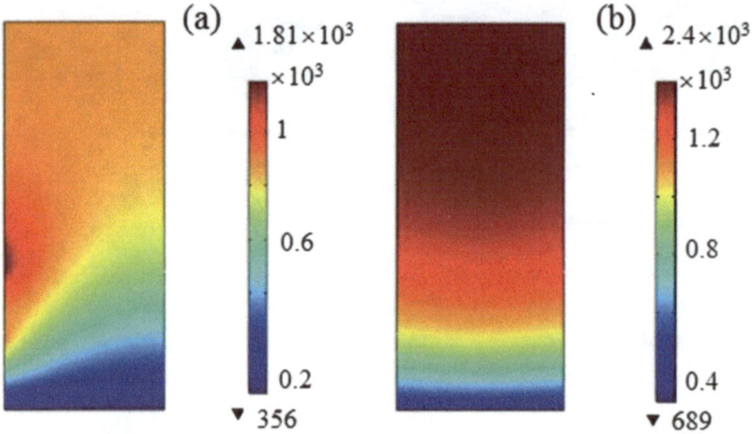

Fig. 3.37 2D distributions of gas temperature (K) under single port (**a**) and dual port (**b**), with coherent power-combining

Nevertheless, by using the coherent power-combining technique, the electrons are distributed symmetrically about the axis of the quartz tube, as shown in Fig. 3.36b. Due to the simultaneous absorption from a dual port, the microwave energy is effectively utilized for discharging. The maximum electron density reaches 2.01×10^{19} m^{-3}, which is one order higher than that for a single port. Moreover, the maximum and minimum values fluctuate within an order of magnitude, which indicate the uniformity of the plasma. The nonuniformity of the gas temperature directly causes the appearance of squeezed plasma. To design the structure of the microwave plasma apparatus more efficiently and to meet the industrial requirements, the heat effect of the gas discharge should be studied. The 2D distribution of the gas temperature is presented in Fig. 3.36. As shown in Fig. 3.36a, for a single port, the phenomenon of high temperature appears on the sidewall of the quartz tube, close to the microwave port. By using the coherent power-combining technique in Fig. 3.37b, the maximum gas temperature reaches 2400 K, which is 600 K higher than for the single port. Moreover, it is distributed symmetrically and uniformly, which can improve the uniformity of the plasma effectively.

2. Ridge waveguide design in plasma

   In the tapered waveguide, the electric field is compressed at the narrow side of the waveguide, but in the ridged waveguide, the electric field is mainly concentrated under the ridge area, where it reaches its maximum [19]. We conduct our calculation in the 3D domain. This simulation requires multiphysics coupling, including electromagnetic wave propagation, electron transport, heat transfer, and gas flow. Changes in the electron density, electron temperature, gas temperature, and scattering parameter S_{11} with time, which manifest the physical process of the microwave discharges, are achieved. Finally, the plasma reaches a steady state. All these parameters in the ridged waveguide are compared with those in the tapered waveguide. Significantly, this ridged waveguide has a greater power

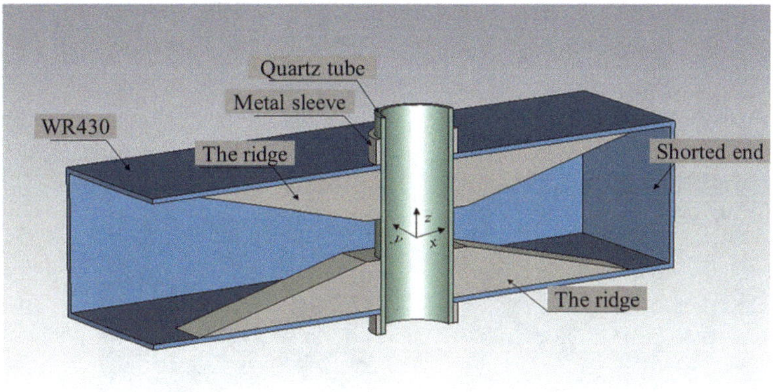

Fig. 3.38 Three-dimensional model of the ridged waveguide

coupling efficiency. Moreover, the plasma in this structure has a higher electron density, electron temperature, and gas temperature at the same input microwave power.

To show the interior of the structure, we cut our actual model into two halves, as shown in Fig. 3.38. It is based on a standard rectangular waveguide, WR430, and is short-circuited at the end. The incident microwave is operated at a frequency of 2.45 GHz. The length of the waveguide is designed to be 218 mm, which is about 1.5 λ_p (λ_p is the waveguide wavelength) in order to gain a higher electric field intensity in the discharge region. Two identical trapezoidal ridges are aligned and embedded on the center of the inner top and bottom surfaces. The thicknesses of the two ridges are 32 mm and their wide and narrow sides are set to 180 mm and 42 mm, respectively. The transitions of the two ridges are smooth, to ensure less reflection of microwave power, and the angle of inclination is optimized to be 16°. In the discharge region, a quartz tube with an inner diameter of 26 mm and an outer diameter of 30 mm penetrates the center of the waveguide's wide walls. Argon at a temperature of 293.15 K and atmospheric pressure passes through the gas inlet into the quartz tube at a rate of 64 L/min. The other spaces are filled with air.

The plasma simulation includes the electromagnetic propagation model, heat transfer model, and fluid flow module, and provides the electric fields, gas temperature, and gas velocity. All of the results can be observed in Figs. 3.39, 3.40, 3.41, and 3.42.

The microwave plasma is significantly affected by the microwave power and microwave frequency. For different industrial applications, the parameters of the plasma are different. Two types of waveguides are modeled to check the effects of frequency variations. The standard rectangular waveguide BJ-9 is used to propagate 915 MHz microwaves and BJ-22 is used to propagate 2450 MHz microwaves. The algorithm proposed in Sect. 4.2.2 was used to calculate the electron density, electron temperature, and gas temperature. The results are shown in Fig. 3.43.

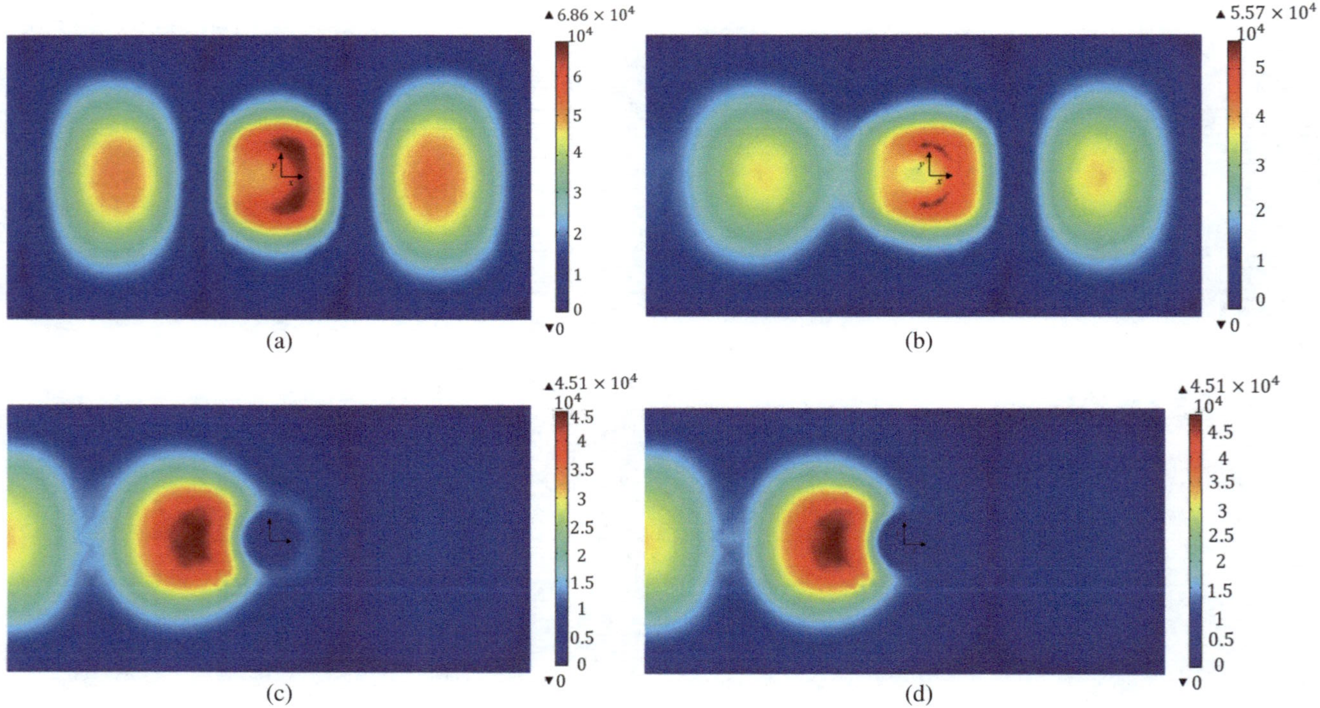

Fig. 3.39 Slice plots of the electric field module (V/m) distributions inside the waveguide in the x-y ($z = 0$) plane and the quartz tube in the x-z ($y = 0$) plane at different moments of the discharge process. (a) 1×10^{-9} s, (b) 3×10^{-6} s, (c) 4×10^{-4} s and (d) 1 s

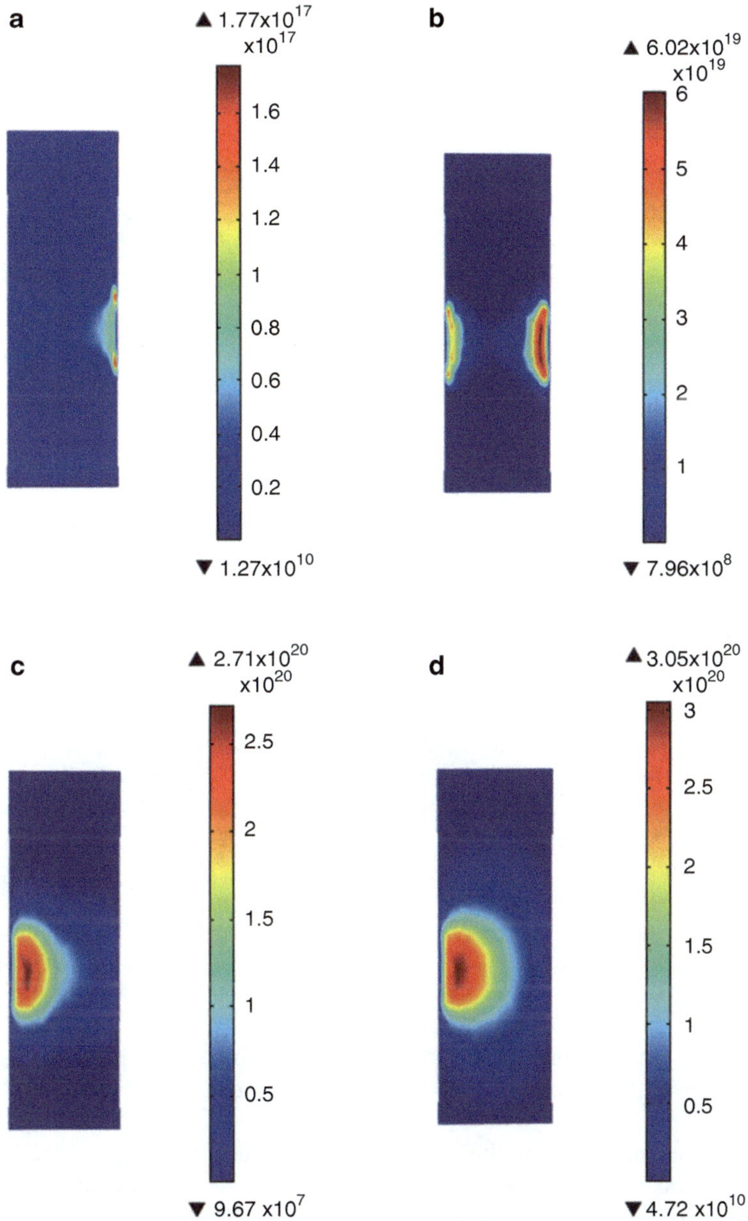

Fig. 3.40 Slice plots of the electron density (m^{-3}) distributions inside the quartz tube in the x-z ($y = 0$) plane. (**a**) 2×10^{-7} s, (**b**) 1×10^{-5} s, (**c**) 1×10^{-4} s, and (**d**) 1 s

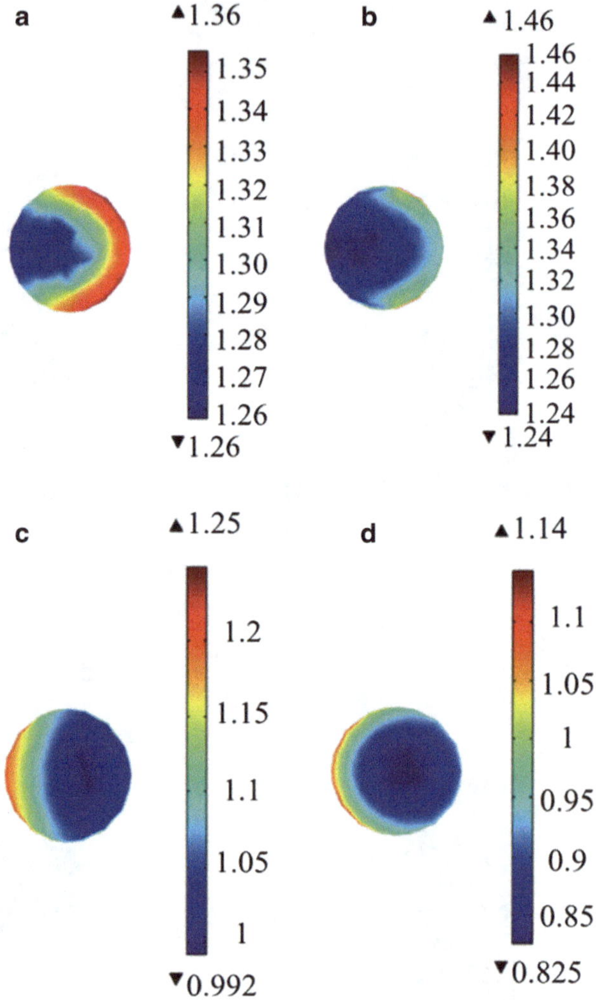

Fig. 3.41 Slice plots of the electron temperature (V) distributions inside the quartz tube in the x-z (y = 0) plane. (**a**) 8×10^{-8} s, (**b**) 1×10^{-6} s, (**c**) 2×10^{-4} s, and (**d**) 1 s

Moreover, the incident microwave power is changed to check the electron density, electron temperature, gas temperature, and torch length. Detailed results are shown in Fig. 3.44.

These plots indicate that the electron density, electron temperature, and gas temperature increase with the input power. However, in the experiments for measuring the torch length, when the input power changes from 1000 W to 2000 W, the torch length increases by just 1.5 mm.

Microwave plasma is a new and important research with many advantages and applications. Here, we have already introduced the simulation methods (multiphysics computation) and structure designs (ridged waveguide and dual-port waveguide). However, there are many other issues we still need to explore, such as the diagnosis of microwave plasma, the applications in industry, and the

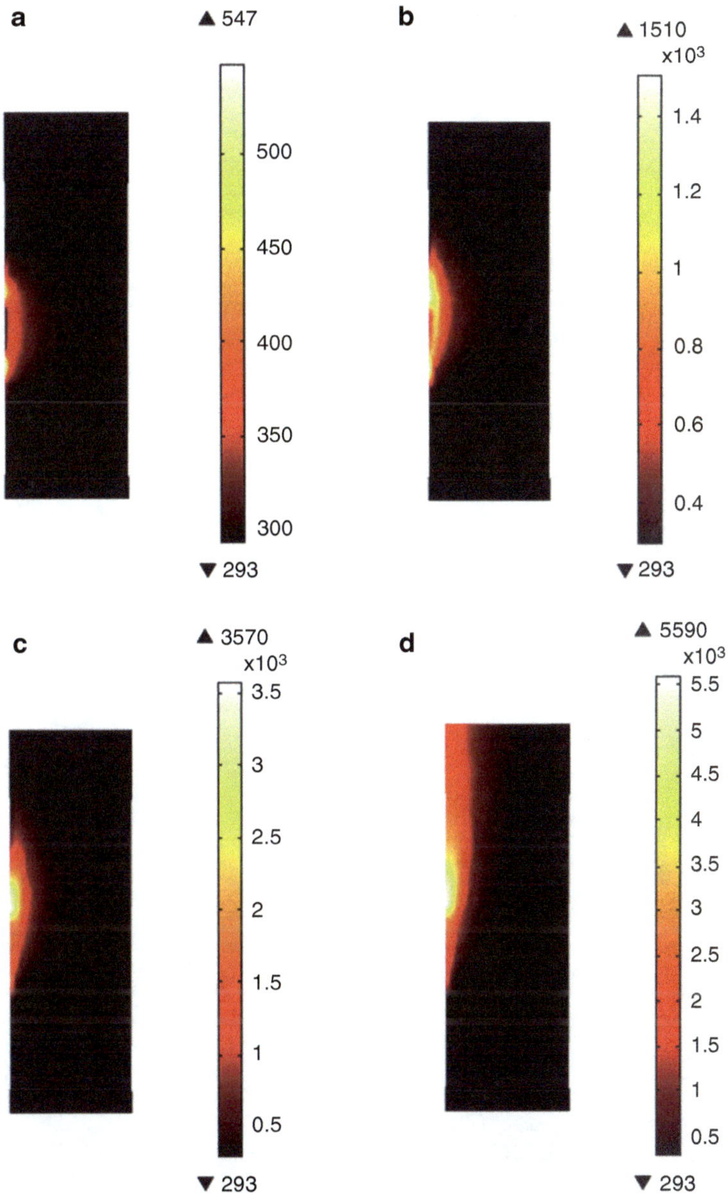

Fig. 3.42 Slice plots of the gas temperature (K) distributions inside the quartz tube in the x-z (y = 0) plane. (**a**) 9×10^{-4} s, (**b**) 0.005 s, (**c**) 0.01 s, and (**d**) 1 s

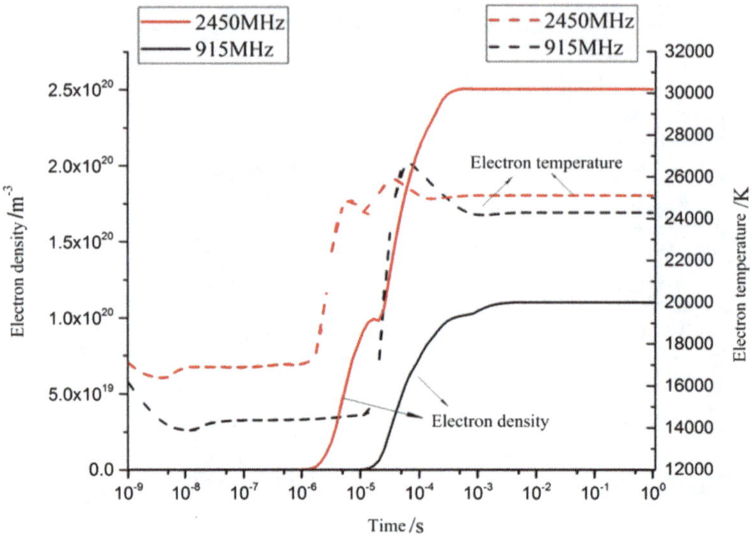

Fig. 3.43 Comparison between 915 MHz and 2450 MHz

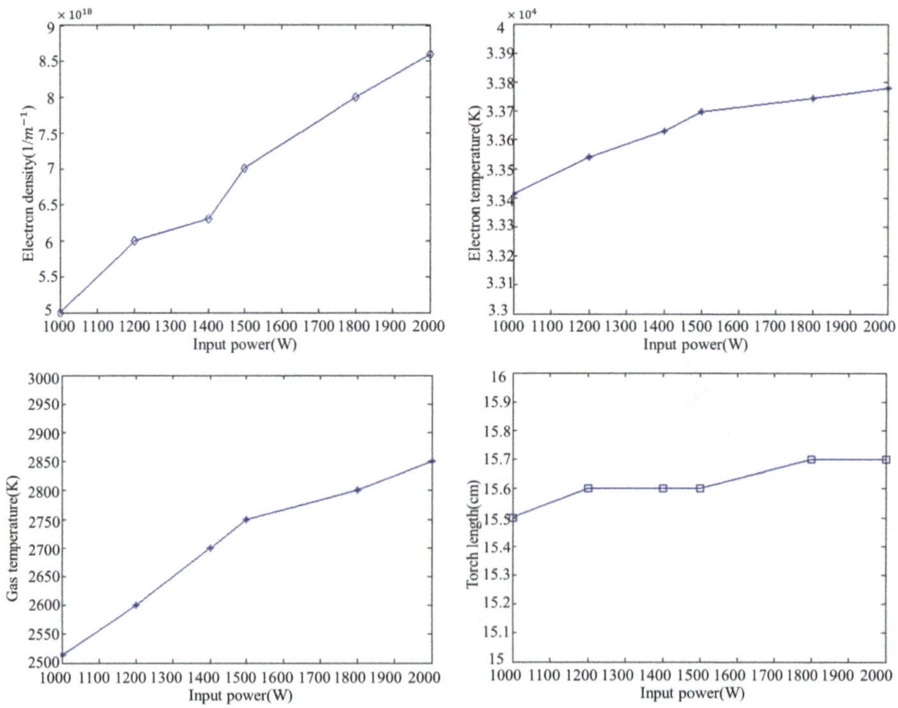

Fig. 3.44 Variation of plasma parameters with input power

optimization of the devices. The deep mechanism of a plasma should also be understood, which would be helpful for industrial applications.

3.5 Summary

In summary, we raise the challenges to microwave energy application, such as nonuniformity heating and thermal runaway, etc. Multiphysics calculation methods was used to overcome these challenges. Based on the basis of dielectric characterization in the previous chapter, the multiphysics models of the complex process under the microwave irradiation is completed. Multiphysics model based on transformation optics algorithm, implicit function and level set methods were built for moving elements during microwave heating, these works were employed to analyze the heating process, and will be useful to design a safety industrial microwave heating equipment.

References

1. George J, Bergman R. Selective re-meshing: a new approach to include mode stirring effects in the steady state FDTD simulation of microwave heating cavities. Microw Opt Techn Lett. 2005;48(6):1179–82.
2. Pozar DM. Microwave engineering. 3rd ed. New York: Wiley; 2005.
3. Ward AJ, Pendry JB. Refraction and geometry in Maxwell's equations. J Mod Optic. 1996;43 (4):773–93.
4. Milton GW, Briane M, Willis JR. On cloaking for elasticity and physical equations with a transformation invariant form. New J Phys. 2006;248(8):1–20.
5. Liao YH, Zhu HC, Tang ZM, Huang KM. Transformation optics for efficient calculation of transmembrane voltage induced on cells. Chin Phys B. 2015;24(10):1002041–5.
6. Dolin LS. To the possibility of comparison of three-dimensional electromagnetic systems with nonuniform anisotropic filling. Radiophys Quantum Electron. 1961;4:964–7.
7. Osher S, Fedkiw R. Level set methods and dynamic implicit surfaces. New York: Springer; 2003.
8. Pozar DM. Microwave engineering. 3rd ed. New York: Wiley; 2005. ch. 1
9. Huang KM, Liao YH. Transient power loss density of electromagnetic pulse in debye media. IEEE Trans Microw Theory Techn. 2015;63(1):135–40.
10. Baeva M, Andrasch M, Ehlbeck J, Loffhagen D, Weltmann K-D. Temporally and spatially resolved characterization of microwave induced argon plasmas: experiment and modeling. J Appl Phys. 2014;115(14):143301-1–143301-13.
11. Vahedi V, Surendra M. A Monte Carlo collision model for the particle-in-cell method: applications to argon and oxygen discharges. Comput Phys Commun. 1995;87:179–98.
12. Kushner MJ. Hybrid modelling of low temperature plasmas for fundamental investigations and equipment design. J Phys D Appl Phys. 2009;42(19):194013-1–194013-20.
13. Van Dijk J, Kroesen GMW, Bogaerts A. Plasma modelling and numerical simulation. J Phys D Appl Phys. 2009;42(19):190301-1–190301-14.

14. Ferreira CM, Loureiro J, Ricard A. Populations in the metastable and the resonance levels of argon and stepwise ionization effects in a low-pressure argon positive column. J Appl Phys. 1985;57(1):82–90.
15. Rogoff GL. Ambipolar diffusion coefficients for discharges in attaching gases. J Phys D Appl Phys. 1985;18(8):1533–45.
16. Li S-Z, Zhang X, Chen C-J, Zhang J, Wang Y-X, Xia G-Q. Role of discharge tube in determination of operating mode in waveguide-based atmospheric-pressure microwave-induced plasma. IEEE Trans Plasma Sci. 2014;42(10):2776–7.
17. B. M. Smirnov. Theory of gas discharge plasma. Moscow: Springer; 2015.
18. Zhao Q, Liu SZ, Tong HH. Plasma technology and its applications. Beijing: National Defence Industry Press; 2009. p. 105.
19. Ruiz-Cruz JA, El Sabbagh MA, Zaki KA, Rebollar JM, Zhang Y. Canonical ridge waveguide filters in LTCC or metallic resonators. IEEE Trans Microw Theory Techn. 2005;53(1):174–82.

Chapter 4
Interaction Between Microwave and Molecules

4.1 Introduction

The previous chapters indicated that microwave is widely used due to their advantages in the chemical industry. However, in actual industry applications, scientists have gradually discovered some unusual phenomena of microwave in chemistry, which make the industrial application of microwave uncertain. This kind of uncertainty may make the microwave in industrial application fail to play the expected role, or even bring safety accidents. We have found the following unusual phenomena in our work: (1) The activation energy and pre-exponential factor of chemical reaction will change when microwave is applied, which will lead to the change of chemical reaction rate. (2) The conductivity of NaCl aqueous solution will change with microwave power. (3) Special complex permittivity of a mixture will be higher or lower than both. (4) Nonlinear effect of dielectric properties of solution with microwave. Herein, we will introduce the four special phenomena of the interaction between microwave and chemicals found in experiments to show the unusual effect of microwave.

4.1.1 Activation Energy and Pre-exponential Factor of Chemical Reactions

To illustrate the mechanism of the effect of microwave on chemical reaction, the fundamental work is to investigate the influence of microwave on the activation energy and pre-exponential factor. As early in 1996 [1], we have investigated the effect of microwaves on the two well-known oxido-reduction reactions of KI with H_2O_2 and $KMnO_4$ with $H_2C_2O_4$.

The variations of ln ($1/t$) (t represent the time) as a function of $1/T$ (T represent the Tempreture) in experiments I and II are shown in Figs. 4.1 and 4.2, respectively.

K. Huang et al., *Dynamics in Microwave Chemistry*,
https://doi.org/10.1007/978-981-15-9655-1_4

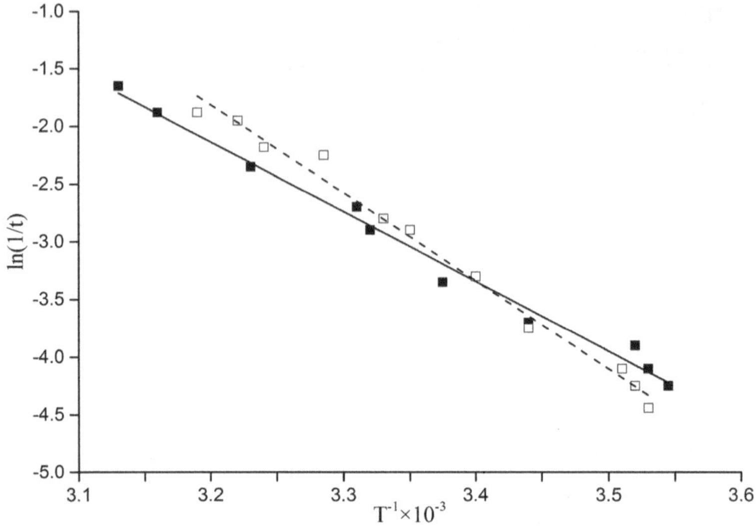

Fig. 4.1 Variation of ln(1/t) as a function of T^{-1} in experiment I. (filled square) control group; (open square) test group

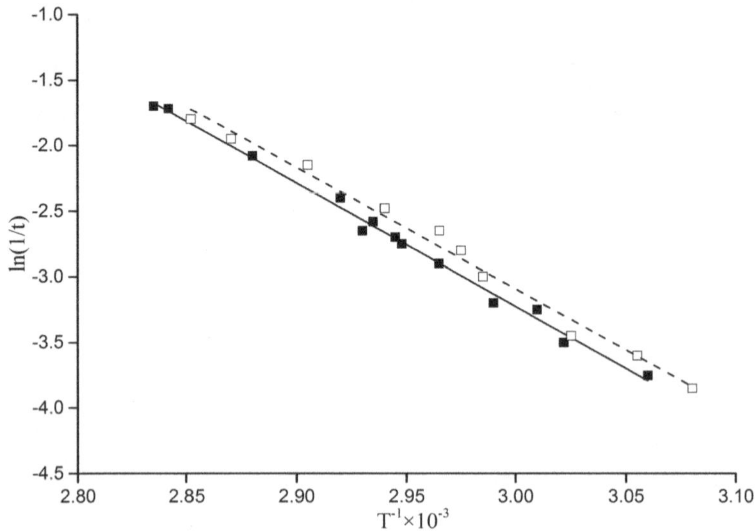

Fig. 4.2 Variation of ln(1/t) as a function of T^{-1} in experiment II. (filled square) control group; (open square) test group

From Figs. 4.1 and 4.2 it can be seen that the relationship between ln(l/*t*) and the reactions without irradiation of electromagnetic waves is suitable for Arrhenius equation, or in other words, it is linear. The linear correlation coefficients are −0.993 for experiment I and − 0.997 for experiment II. The activation energy of the reactions can be calculated from Fig. 4.1 and 4.2. The activation energy obtained from Fig. 4.1 is $\Delta E = 54.6$ kJ/mol, while the activation energy obtained from Fig. 4.2 is $\Delta E = 79.08$ kJ/mol. The interesting result is that the relationship between ln(l/t) and l/T of reactions irradiated by electromagnetic waves is obviously nonlinear. The correlation coefficients of the curves in Figs. 4.1 and 4.2 are -0.998 and $-$ 0.993, respectively.

In recent years, many studies have reported that microwave can change activation energy and pre-exponential factor of chemical reactions. A comparative study between the conventional and 2.45 GHz microwave multimode sintering behavior of insulator (α-Al_2O_3) and semi-conductive ceramic (ZnO) was systematically investigated by Zuo et al. [2]. During microwave densification process, the apparent activation energy was about 90 kJ/mol less than the value for conventional sintering of Al_2O_3. However, an opposite result was obtained in the case of ZnO, although its densification process had been also accelerated by microwave as well as Al_2O_3 in Fig. 4.3. Mazo et al. [3] have studied the self-esterification of partially maleated castor oil using conventional and microwave heating. This reaction proceeded more

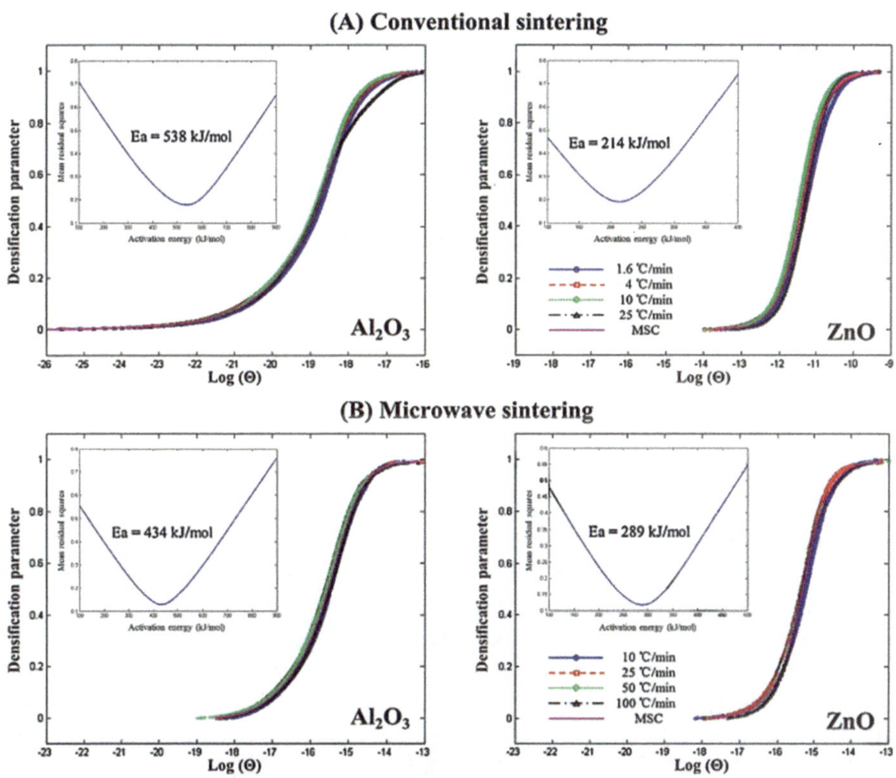

Fig. 4.3 Master Sintering Curves under conventional (**a**) and microwave (**b**) sintering of Al_2O_3 and ZnO for the whole stage

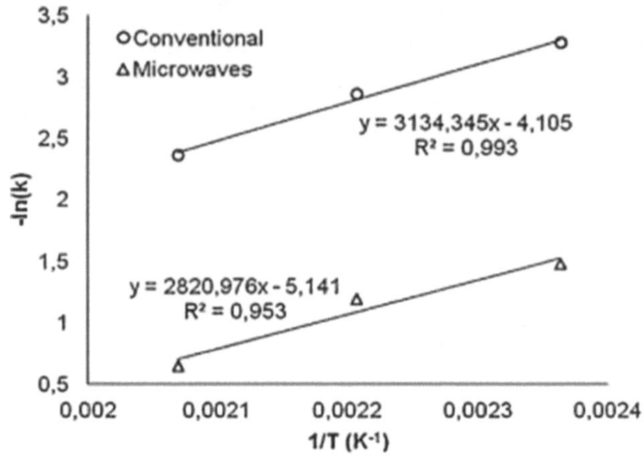

Fig. 4.4 Plotting of the Arrhenius equation, using conventional and microwave heating

quickly when microwave was used and showed the presence of non-thermal effects of microwave heating, which were evidenced by the 10% decrease in the activation energy and the 182% increase in the pre-exponential factor in Fig. 4.4. The CO_2 gasification of oil palm shell char to produce CO through the Boudouard reaction was investigated under microwave irradiation by Lahijani et al. [4]. According to Fig 4.5, these results showed a decline of six orders of magnitude in the pre-exponential factor with microwaves. These investigations of the variation of pre-exponential factor and activation energy of reactions irradiated by microwave are very important for illustrating the principle of the effect of microwave on chemical reaction and efficient application of microwave to chemical industry.

4.1.2 Conductivity of NaCl Aqueous Solution Changes with Microwave

We have designed an experiment for corroborating the existence of a microwave non-thermal effect on electrolyte aqueous solutions in low-level intensity electric fields [5]. A specially designed ridged waveguide was used to produce a uniform electric field, and the electric field intensity in solution was about 10^4 V/m. Next, we employed a circuit to measure the slight electrical conductivity (EC) variations of a solution flowing through the ridge gap under microwave irradiation, as shown in Fig. 4.6. Some special techniques, such as precise temperature measurements and multi-physics calculations including the electromagnetic field, fluid dynamics and heat transfer, were used to exclude the temperature effect. By measurement, slight EC variations with microwave power at 2.45 GHz were observed and explored. The results reveal the existence of a non-thermal microwave effect.

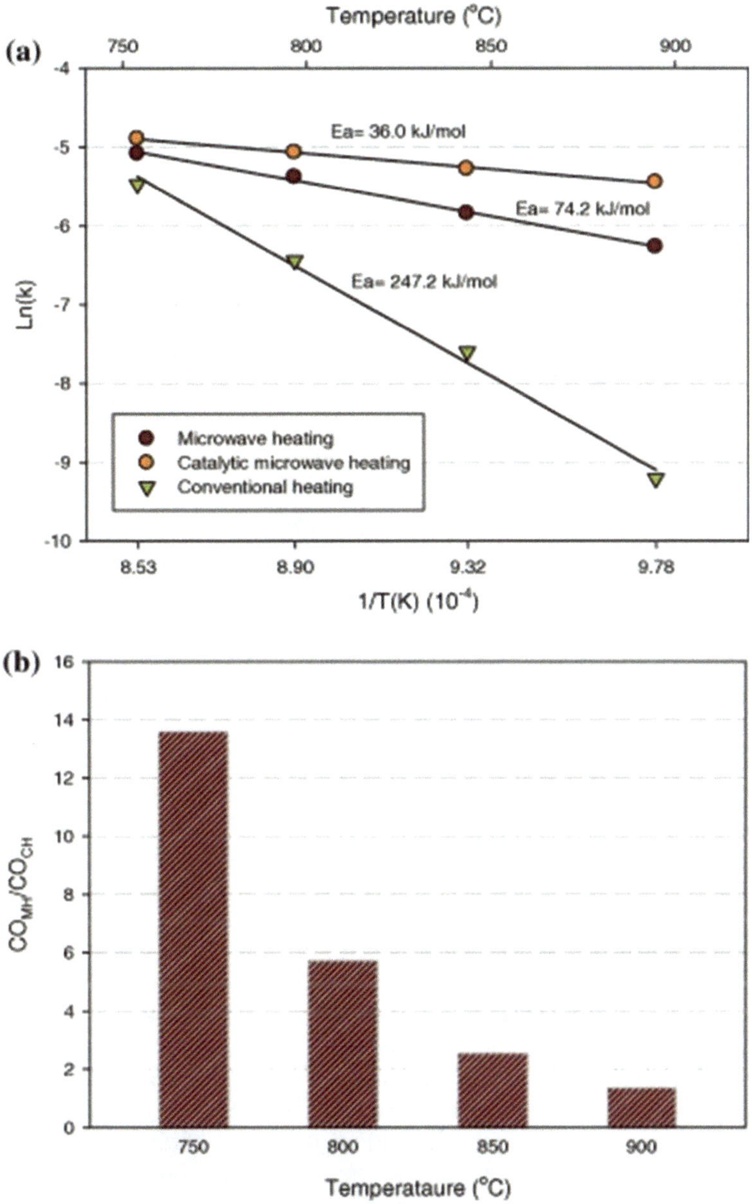

Fig. 4.5 (**a**) Arrhenius plot for catalytic and non-catalytic microwave and thermal driven Boudouard reaction and (**b**) production of CO in microwave (MH) over thermal heating (CH) after 60 min

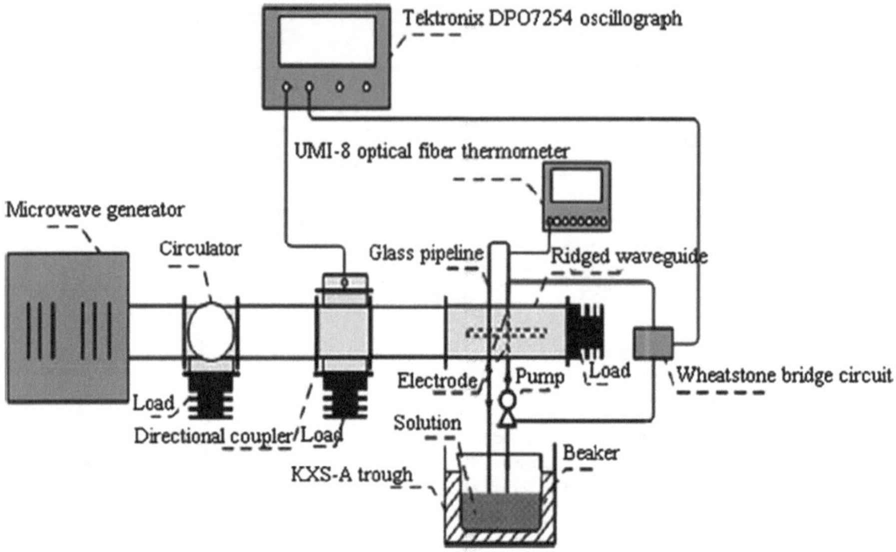

Fig. 4.6 The experiment system

The precisely designed experiments included some special techniques: (1) A specially designed ridged waveguide with a hole in the side wall was used to produce a uniform distribution electric field in the solution; the solution was pumped to flow through the ridge gap in a quartz glass pipeline with a diameter of 5 mm. The space between a pair of ridges was 10 mm, the electric field intensity in the solution was in the order of 10^4 V/m and the input power was 400 W. (2) The temperature of the solution under test was precisely controlled by using a KXS-A trough (± 0.5 °C) and pump (5 m/s). (3) The temperature at the output port of the pipeline was precisely measured by using a UMI-8 optical fiber thermometer with a 1 mm diameter optical fiber. (4) A Wheatstone bridge circuit was used to accurately measure the voltage variations resulted from the slight EC changes of the solution. (5) To observe the concomitant EC change with microwave power, the microwave power was on–off-modulated irregularly during the experiments.

The measured EC variations with microwave power of sodium chloride aqueous solutions at three constant temperatures are shown in Fig. 4.7. It is obvious that the EC of the sodium chloride aqueous solutions depended on the microwave power directly. The higher the concentration of solution, the more obvious the EC change. By measurement, the slight EC variations of sodium chloride aqueous solutions with microwave power validate the existence of a non-thermal microwave effect in low-level intensity electric fields. However, this effect is too weak to be detected in typical applications of microwave. More experiments under the different conditions, such as different frequencies and electrolytes, are required to verify this effect further. We assume that the effect results from cluster structure changes in electrolyte

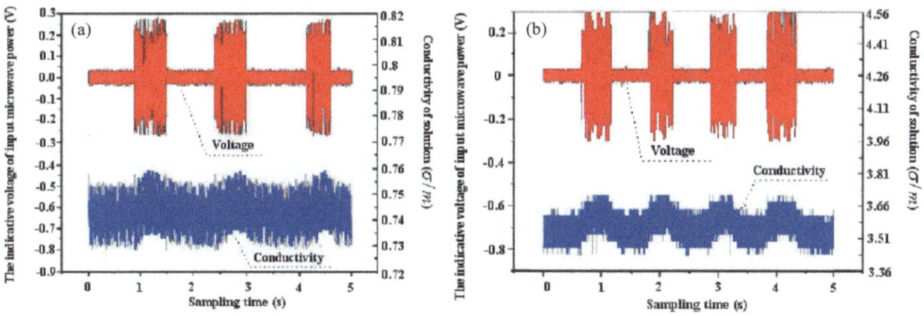

Fig. 4.7 The EC variations with microwave power of sodium chloride aqueous solutions at three constant temperatures: (**a**) 0.1 mol/L sodium chloride aqueous solution at 10 °C, (**b**) 0.5 mol/L sodium chloride aqueous solution at 10 °C

aqueous solutions under microwave irradiation, but evidence of this requires further study.

4.1.3 Special Permittivity of Mixed Solution

To study the dielectric behavior of a mixed system, Wu et al. measured the permittivity of pyridine–ethanol mixed solutions by using a vector network analyzer (ZVA50) and a DAKS-3.5 probe, as shown schematically in Fig. 4.8. The solutions used in these experiments included ethanol (99.5%, analytical reagent) and pyridine (99.5%, analytical reagent) (purchased from Chengdu Kelong Chemical Reagent Factory), and all of them were used as purchased without further purification. All pyridine–ethanol mixed solutions (the volume fraction of pyridine V(pyridine)% was 0–100%) were prepared. Every sample was measured at least four times on different occasions, and the measurement error was maintained within 5%.

At 2.45 GHz, the dielectric spectrum of the pyridine–ethanol mixed solutions at 293.15, 303.15, and 313.15 K are plotted in Fig. 4.9a–c, respectively. Figure 4.9 shows that the real part, ε_r', of the permittivity tends to increase first and then decrease at the three different temperatures as the fraction of pyridine increases. Surprisingly, we observe an exceptional behavior of the real part, ε_r', of the mixed solution permittivity, which is larger than those of their pure components at certain concentrations. Likewise, the imaginary part, ε_r'', has the same inverted U-shaped tendency with increasing fraction of pyridine at 293.15 and 303.15 K. The unusual phenomenon that the imaginary part, ε_r'', is larger than those of their pure components at certain concentrations also appears. However, when the temperature increases to 313.15 K, the exceptional behavior of ε_r'' disappears.

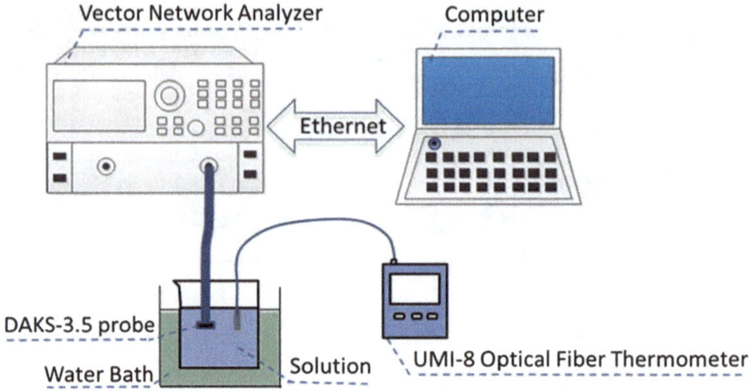

Fig. 4.8 Schematic of the experimental system

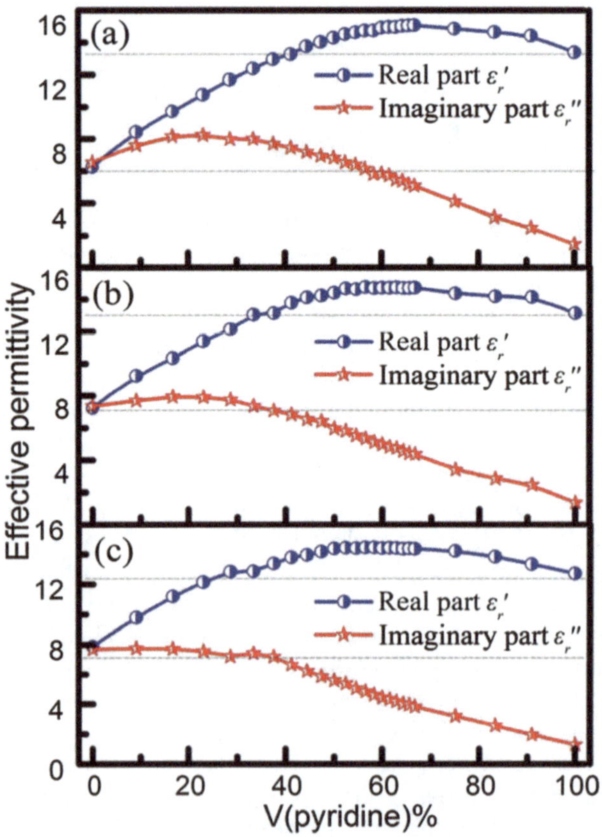

Fig. 4.9 Permittivity of pyridine–ethanol mixed solutions at 2.45 GHz as a function of the volume fraction of pyridine V(pyridine)%. (**a**) 293.15 K, (**b**) 303.15 K, (**c**) 313.15 K

4.1.4 Nonlinear Dielectric Properties with Microwave

To study the relation between dielectric behavior and microwave power, Sun et al. designed an experimental system to measure the dielectric behavior of DMSO-methanol/ethanol mixtures under low-intensity microwave fields (10^5 V/m electric field amplitude). A specially designed microstrip sensor with high sensitivity to dielectric changes was employed to generate a uniform electric field. Then, this highly sensitive sensor was applied to a microwave system to measure the dielectric changes in mixtures that flowed through the designed sensor under microwave irradiation.

The temperature effects caused by microwave heating were excluded by the well-designed device. The designed center frequency is 2.45 GHz, and a picture of the sensor is shown in Fig. 4.10. Supposing that the solution's dielectric properties in the two channels remain unchanged, the two signals will cancel each other. Otherwise, if the dielectric properties of the solution in the MUT channel changes, their difference is recorded.

Pure DMSO (99.5%), methanol (99.5%), and ethanol (99.5%) (analytical reagent) solutions (purchased from Chengdu Kelong Chemical Reagent Factory) were used without further purification in these experiments. Two sets of primary alcohol–DMSO mixtures, methanol–DMSO (methanol mole fractions 1.75, 0.59, and 0.35) and ethanol–DMSO (ethanol mole fractions 1.22, 0.41, and 0.24), were prepared. In the experiments, a continuous microwave source was employed, and the generator's power was ranged from 20 W to 100 W, operating at a frequency of 2.45 GHz. When the input power was gradually increased from 20 to 100 W, the measured ratio values between the output and input power were plotted in the

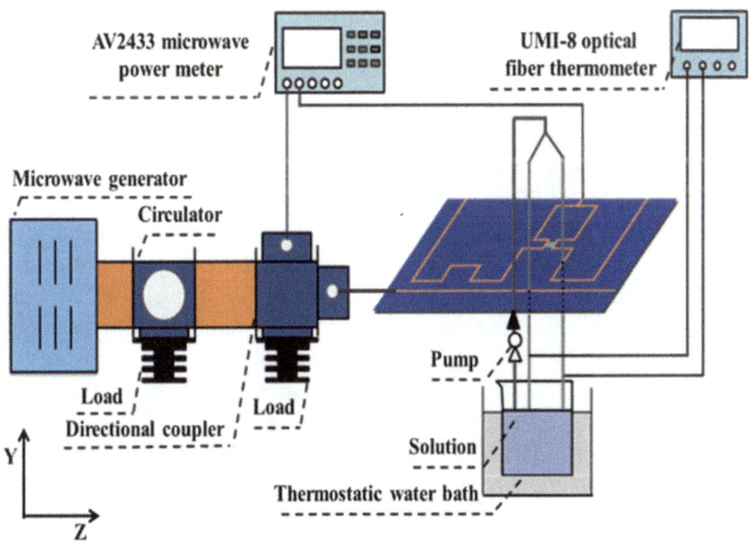

Fig. 4.10 A schematic of the experimental system

Cartesian coordinate system. If the ratio curves were not linear, it meant that the dielectric property of the solution in the MUT channel had changed. Each binary mixture was measured at a room temperature of 300 K, and repeated six times to eliminate the systematic errors. The results (with error bars) are shown in Fig. 4.11.

As illustrated in Fig. 4.11a, b, similar characteristics were found in the curves of the methanol−DMSO and ethanol−DMSO mixtures with different mole fractions. The output power versus input power curves were initially linear, and then, showed nonlinear characteristics. The curve's slope became smaller at some "critical point," which can be viewed as evidence for a change in the electric field-induced dielectric property of these mixtures. The "critical point" is defined as the value of the input power corresponding to the point where the nonlinear characteristic first appeared. The results are shown in Fig. 4.11c, d. These measured linear results further excluded the trivial effects caused by the high-power microwaves (such as electromagnetic crosstalk), and proved the reliability of the experimental system. In summary, by comparing the mixtures' nonlinearity with the pure substances' linearity, it can be deduced that low-intensity microwave can affect the inter-molecular interaction in DMSO–methanol/ethanol mixtures, and this microscopic interaction eventually results in macroscopic dielectric changes in the measurements under the stimulus of a 10^5 V/m electric field.

The results validate the existence of a special microwave effect in low-level-intensity electric fields. Further theoretical work on the details of the interaction mechanism between the microwaves and mixtures should be studied from the molecular level.

These special phenomena cannot be explained by the thermal effect of microwave. However, the photon energy of microwaves is not strong enough to disrupt chemical bonds, so the special effect cannot be explained by quantum theory either. Thus, the special effects of microwave become a great scientific obstacle in applications. To interpret these special phenomena in principle, it is necessary to study the interaction between microwave and matter at the molecular level. The molecular dynamics (MD) method, as a computer method at the molecular level, can well meet the needs of this research.

In this chapter, the MD method has been successfully applied to mixed molecular solutions and an ionic aqueous solution, which agree well with the experimental phenomenon. For example, by studying the dielectric properties of binary solvent mixtures of dimethyl sulfoxide with water, the static complex permittivity, the dielectric relaxation at the frequency of microwaves, as well as the microwave spectra of DMSO−water mixtures over the whole composition range at room conditions, have been investigated by MD simulations in order to research the special phenomena of the mixture solution's dielectric properties. The studies completed in this chapter indicate that the MD method has a great application prospect in the field of microwave chemistry.

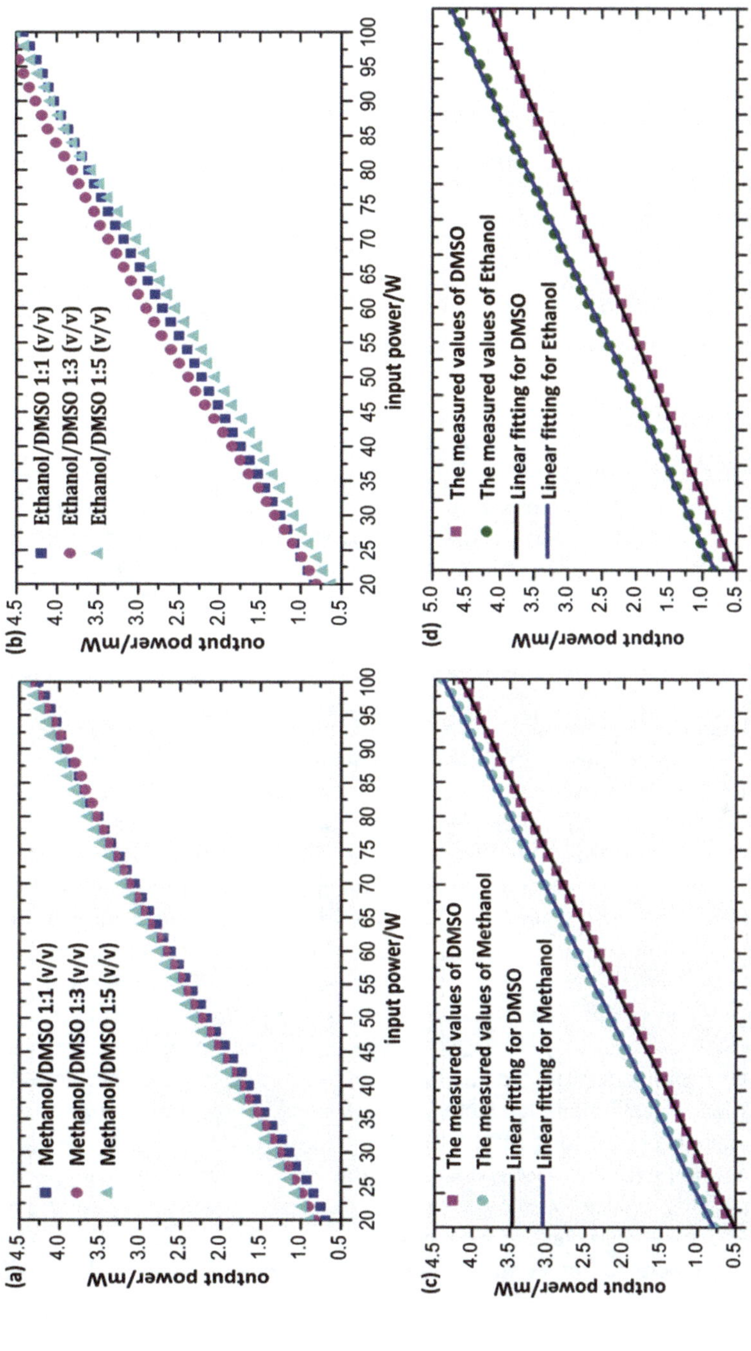

Fig. 4.11 Relations of output−input microwave power for solutions of different mole fraction ratios. (**a**) Methanol/DMSO mixtures at methanol mole fractions of 1.75, 0.59, and 0.35. (**b**) Ethanol/DMSO mixtures at ethanol mole fractions of 1.22, 0.41, and 0.24. (**c**) Pure DMSO (99.5%) and methanol (99.5%) solutions. (**d**) Pure DMSO and ethanol (99.5%) solutions

4.2 MD Simulation

4.2.1 Introduction

Scientists have been trying to study physical systems at the atomic or molecular level. Since the rapid development of quantum theory in the twentieth century, nearly all properties of a molecule, such as structure, conformation, dipole moment, ionization potential, and electron density, can be obtained by quantum mechanics calculations, and are found to agree well with the experimental results. However, the quantum method applies only to some simple molecules or a system with a few electrons. Most systems in nature or in industry, such as biochemical molecules (proteins, enzymes) or polymers, contain several atoms and electrons, which makes calculations difficult. For systems of metallic materials, polymeric materials, solution, solid mixtures, or nanomaterials, the properties of a single molecule or the interaction between two molecules are not as important as the properties of the set, dynamic behavior, and thermodynamic properties. To solve the problem of a "huge" system, scientists set out to research various feasible nonquantum computing methods since the 1960s and made substantial progress. Then, the MD method and the Monte Carlo method were developed. MD is a computer simulation method for studying the physical movements of atoms and molecules, and is thus a type of N-body simulation. Compared to the Monte Carlo method, MD has a higher accuracy and validity in the calculation of macroscopic properties, so it can be widely used in physics, chemistry, biology, materials science, medicine, and other fields.

Here, we introduce a classical, efficient molecular research method, MD, and described how it can be used in microwave chemistry.

MD is a dominant approach that has been most widely adopted to compute large and complex systems, as shown in Fig. 4.12. Since 1970, many force fields for biochemical systems, polymers, metal, and nonmetallic materials have been systematically established because of the rapid development in molecular mechanics, which has significantly enhanced the ability to compute the structure, spectroscopic properties, and thermodynamic properties of materials. Compared to the Monte Carlo method, the MD method has a more correct physical basis as the force field and adopts Newton's laws of motion. Now, scientists have improved the computational techniques of MD and many commercial software based on MD have been released, with strong computing power and the ability to deal with various problems. In advanced laboratories at a school, a factory, or a hospital, these commercial software applications have become indispensable for research.

At the beginning of the 1990s, the algorithm of molecular simulations also evolved to handle a real molecular system with the development of the force field. The molecule's force field can depict a simple system and includes most types of molecules. Since then, computer simulations have opened a new era in the simulation field.

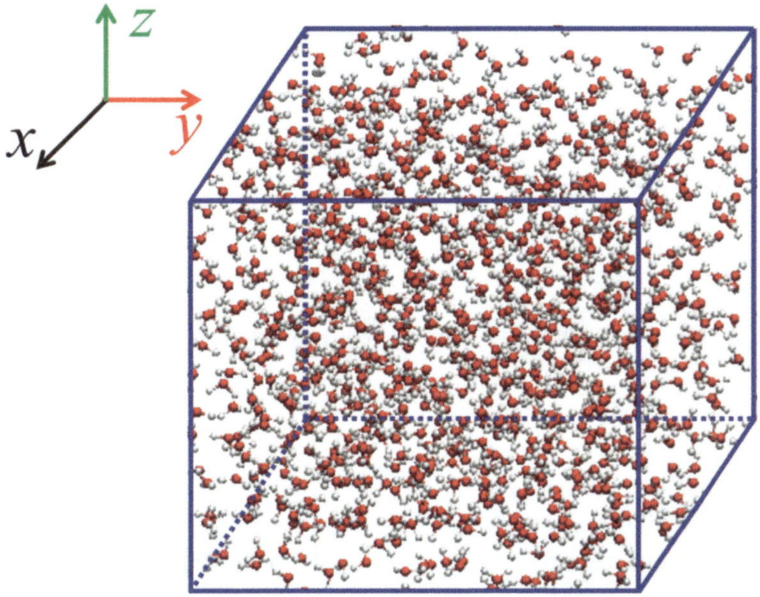

Fig. 4.12 Complex system with 1000 molecules

4.2.2 MD in Electrostatic Fields

The MD simulation method is very similar to the real experiment. In an experiment, the following three steps are required: preparing the sample, measuring the sample by using the test instrument, and analyzing the experimental results. The MD simulation method follows similar steps: (1) preparing the sample, namely the establishment of a model system by N molecules/particles (usually based on the Maxwell–Boltzmann velocity distribution), (2) data collection, a solution of the Newtonian equation included by the N molecules/particles model system until it balances, and a calculation of its properties, and (3) result analysis.

There is nothing special about the MD simulation under an electrostatic field, compared to a classical MD simulation, except that an electric field parameter is included in the simulation. Therefore, their simulation process is consistent, as described below.

If there are N molecules in the system, the energy of the system is the sum of their kinetic energy and potential energy. According to the Maxwell-Boltzmann velocity distribution, the initial velocity of each molecule will be given. Then, their kinetic energy, E_k, can be represented by

$$E_k = \sum_{i=1}^{N} \frac{1}{2} m_i \vec{v}_i^2 \tag{4.1}$$

Here, m_i and $\vec{v}_i$ indicate the mass and velocity of the ith molecule, respectively. The temperature of the system is given by the total kinetic energy of the N molecule system. For the absolute temperature T, the total kinetic energy can be calculated by

$$E_k = \frac{1}{2} N_{df} K_B T \tag{4.2}$$

where K_B is the Boltzmann constant and N_{df} is the number of degrees of freedom, which can be computed as

$$N_{df} = 3N - N_c - N_{com} \tag{4.3}$$

Here, N_c is the number of constraints imposed on the system and N_{com} is the number of combinations.

The total potential energy $V\left(\vec{r}_1, \vec{r}_2, \ldots, \vec{r}_n\right)$ is a function of the position of each atom in the system, and can be divided into two parts: the intermolecular (or interatomic in molecules) non-bond interactions (Van der Waals interactions, V_{vdw}) and the intramolecular potential energies (V_{int}).

$$V = V_{vdw} + V_{\text{int}} \tag{4.4}$$

In general, the Van der Waals interactions, V_{vdw}, can be approximately represented by the sum of the Van der Waals interactions of each atom pair.

$$V_{vdw} = V_{12} + V_{13} + \cdots + V_{1N} + V_{23} + V_{24} + \cdots = \sum_{i=1}^{n-1} \sum_{j=i+1}^{n} V_{ij}\left(r_{ij}\right) \tag{4.5}$$

Here, r_{ij} represents the distance between the ith and jth atoms.

Based on the classical mechanics, the force acting on the ith atom in the system can be represented as the gradient of the potential energy ($V_{\text{potential}}$):

$$\vec{F_i} = -\nabla_i V = -(e_x \frac{\partial}{\partial x_i} + e_y \frac{\partial}{\partial y_i} + e_z \frac{\partial}{\partial z_i}) V \tag{4.6}$$

In addition, the motion of each molecule obeys Newton's equations:

$$m_i \ddot{\vec{r}}_i(t) = \vec{F}_i(t) \tag{4.7}$$

where $\vec{r_i}(t)$ represents the positions of the ith molecules. $\vec{F}_i(t)$ represents the forces acting on the ith molecule. By integrating Eq. (4.7) with time t, the position $\vec{r}_i(t)$, velocity $\vec{v}_i(t)$, and acceleration $\vec{a}_i(t)$ of the ith molecule at time t can be obtained as follows:

$$\vec{r}_i(t) = \vec{r}_i(0) + \vec{v}_i(0)t + \frac{1}{2}\vec{a}_i(t)t^2\vec{v}_i(t) = \vec{v}_i(0) + \vec{a}_i(t)t\vec{a}_i(t) = \ddot{\vec{r}}_i(t) \quad (4.8)$$

Here, $\vec{r}_i(0)$ and $\vec{v}_i(0)$ are the initial velocity and initial position of the ith molecule, respectively. Now, the key issue is to solve Newton's equations of motion for a system of N interacting molecules, which is also the basic idea of MD simulations.

To further calculate the thermodynamic quantities and other macroscopic properties of the system, it is necessary to integrate the system. Thus, the integral configuration is very important. In the past 30 years, many effective integral algorithms have been proposed, such as the Verlet algorithm, the leap-frog algorithm, the velocity Verlet algorithm, and the gear predictor-corrector method.

4.2.3 *MD in Microwave Fields*

As we all know, the electric and magnetic fields that make up a microwave are perpendicular to each other, and the direction of the microwave is determined by the $\vec{E} \times \vec{H}$ (as presented in Fig. 4.13).

That is to say that if electric field is in the X direction and the magnetic field is in the Y direction, the microwave propagates in the Z direction. And because of the phase difference ($\frac{\pi}{2}$) between the electric field and the magnetic field, the applied $\vec{E}$(t) and $\vec{H}$(t) can be defined as:

$$\begin{aligned} \vec{E}(t) &= E_{\max} \cos(\omega t)\hat{e}_x \\ \vec{H}(t) &= H_{\max} \sin(\omega t)\hat{e}_y \end{aligned} \quad (4.9)$$

where ω is the angular frequency of the alternating E/M fields. To make it easier to calculate the force, we use the magnetic induction intensity $\vec{B}$ instead of magnetic field $\vec{H}$ in MD. Then Eq. (4.9) can be expressed as

$$\begin{aligned} \vec{E}(t) &= E_{\max} \cos(\omega t)\hat{e}_x \\ \vec{B}(t) &= B_{\max} \sin(\omega t)\hat{e}_y \end{aligned} \quad (4.10)$$

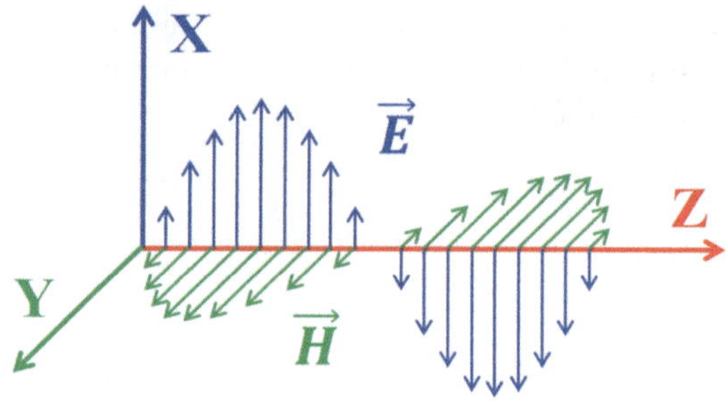

Fig. 4.13 The direction of microwave propagation

The basic principle of molecular dynamics simulation is to use Newton's classical mechanics to solve the trajectory of molecules or atoms in the system. Under microwave, the force acts on the *i*th atom in the system can be written as

$$\vec{F}_{i2} = q_i\vec{E}(t) + q_i\vec{v}_i \times \vec{B}(t) \tag{4.11}$$

The motions of each molecule obey Newton's equations and can be revised as

$$\begin{aligned} m_i\ddot{\vec{r}}_i &= \vec{F}_{i1} + \vec{F}_{i2} \\ &= \vec{F}_{i1} + q_i\vec{E}(t) + q_i\vec{v}_i \times \vec{B}(t) \end{aligned} \tag{4.12}$$

where $\vec{F}_{i1}$ is the force of the *i*th atom when there is no external field. In other words, the force is produced in the central force field produced by the other atoms in the system. The $q_i\vec{E}(t)$ and $q_i\vec{v}_i \times \vec{B}(t)$ are the forces in the electric fields and magnetic fields, respectively. After calculating the force and acceleration of each atom in the system, integrate the equations of Newton's law of motion over Δt to obtain the position and velocity of each atom after passing through Δt. Repeat the steps above, calculate the potential energy of the system from the new position, calculate the force and acceleration of each atom, predict the position and velocity of each molecule after passing Δt. In this way, the position, velocity and acceleration of the molecules in the system can be obtained. In MD simulations, there are many effective integral algorithms to solve Newton's equations of motion. Here, we won't go into details.

4.2.4 MD Simulation in Microwave Fields by GROMACS

Compared with experiments, theoretical studies are of paramount importance, which can reveal the essence of microwave chemical reaction mechanism and is safer and less expensive. In order to perform more suitable molecular dynamics simulation of microwave chemistry, we rewrite the source program of GROMACS software, which is free and open source, to make it available to perform molecular dynamics simulations under external electric, magnetic and microwave fields. The main program of GROMACS software we have redefined is as follows:

```
typedef struct{
  int   g;    /*Number of terms
  real  *d;    /*Coeffients
  real  *rho;  /*Phase angles
        } t_sines;
typedef struct{
  real  B0;    /*
  real  mu;    /*
  real  t0;    /*
  real  tau;    /*
        } t_mfield;


static void calc_f_ma(FILE *fp, int start, int homenr, real charge[],
rvec f[], t_state *state,
            t_sines Bx[], t_sines Bt[], double t)
{
  rvec Bxt;
  real t0;
  int i, m;

  for (m = 0; (m < DIM); m++)
  {
    if (Bt[m].g > 0)
    {
      if (Bt[m].g == 3)
      {
        t0 = Bt[m].d[1];
        Bxt[m] = sin(Bt[m].d[0]*(t-t0))*exp(-sqr(t-t0)/(2.0*sqr(Bt
[m].d[2])));
      }
      else
      {
        Bxt[m] = sin(Bt[m].d[0]*t);
      }
    }
    else
    {
      Bxt[m] = 1.0;
    }
    if (Bx[m].g > 0)
    {
      Bxt[m] *= Bx[m].d[0]*FIELDFAC;
      for (i = start; (i < start+homenr); i++)
      {
       if (m == XX)
        {
         f[i][YY] += charge[i]*state->v[i][ZZ]*Bxt[XX];
```

```
            f[i][ZZ] += (-charge[i])*state->v[i][YY]*Bxt[XX];
          }
          else
          {
            if (m == YY)
            {
              f[i][XX] += (-charge[i])*state->v[i][ZZ]*Bxt[YY];
              f[i][ZZ] += charge[i]*state->v[i][XX]*Bxt[YY];
            }
            else
            {
              f[i][XX] += charge[i]*state->v[i][YY]*Bxt[ZZ];
              f[i][YY] += (-charge[i])*state->v[i][XX]*Bxt[ZZ];
            }
          }
        }
      }
      else
      {
        Bxt[m] = 0;
      }
    }
    if (fp != NULL)
    {
      fprintf(fp, "%10g %10g %10g %10g #MFIELD\n", t,
          Bxt[XX]/FIELDFAC, Bxt[YY]/FIELDFAC, Bxt[ZZ]/FIELDFAC);
    }
}
```

The meaning of most program variables should be apparent from their names, with the same being true for functions. Here, we defined three types of magnetic induction intensity $\vec{B}$ (to make it easier to calculate the force, as described above): static, alternating and pulsing magnetic induction intensity. The alternating magnetic induction intensity is defined as sinusoidal function respect to the cosine alternating electric field. Then, this modified program can be applied to perform molecular dynamics simulations under external electric, magnetic and microwave fields. If you want to use a microwave field in the Z direction, you should set a cosine electric field in the X direction and a sinusoidal alternating magnetic field in the Y direction.

For the sinusoidal alternating magnetic induction intensity, enter three numbers after the B-y: the first number is 1; the second number is the maximal magnetic induction intensity in 1×10^6 T; the third number is the phase of the sine, you can enter a 0 here since the sine of frequency zero has no phase. Then, enter another three numbers after the B-yt: the first number is also 1; the second number is the angular frequency ω in/ps, the third number can also enter any number like 0.

In the same way, there are also six numbers to enter to set a cosine electric field in the X direction. First, enter 3 numbers after the E-x: the first number is 1; the second number is the strength of the electric field in V/nm; the third number is 0. Then, enter another 3 numbers after the E-xt: the first number and the third number are also 1 and 0, respectively. The second number is the angular frequency ω in/ps.

Fig. 4.14 Part of the parameter file

E-x	=	1	1e-5	0
E-xt	=	1	0.015386	0
B-y	=	1	3e-10	0
B-yt	=	1	0.015386	0

For example, we use the modified program to investigate the heating process of water under a microwave field of 2.45 GHz and the electric field of 1×10^4 V/m. The corresponding values of magnetic induction strength are calculated by the wave impedance η and permeability of the medium μ, which are determined as follows:

$$\begin{aligned} \frac{E_{\max}}{H_{\max}} &= \eta \\ \vec{B}(\mathrm{t}) &= \mu \vec{H}(\mathrm{t}) \end{aligned} \tag{4.13}$$

where

$$\begin{aligned} \eta &= \sqrt{\frac{\mu}{\varepsilon}} \\ \varepsilon &= \varepsilon_0 \varepsilon_r \\ \mu &= \mu_0 \mu_r \end{aligned} \tag{4.14}$$

As for pure water, the $\mu = 4\pi \times 10^{-7} H/m$, and the $\varepsilon = 7.168 \times 10^{-10} F/m$. Then, the corresponding value of magnetic induction strength is 3×10^{-4} T with the same frequency of 2.45 GHz. Here, we need to change the frequency 2.45 GHz to angular frequency ω as 0.015386 /ps. According to the description above, the corresponding parameter to set this external microwave field is presented in the Fig. 4.14. As shown in Fig. 4.14, the 1e-5 and 3e-10 are set to express the field intensities of 1×10^4 V/m and 3×10^{-4} T, respectively. And 0.015386 is the frequency of 2.45 GHz.

4.3 Simulation of the Interaction Between Microwave and Molecules

In the previous part, we described the special phenomena of microwave through experiments, and gave the technical means to study these phenomena, namely molecular dynamics simulation. In particular, how to realize the molecular dynamics simulation under microwave field in GROMACS software platform is described. Using this method, we have simulated the molecular dynamics of many chemical systems in microwave field, and achieved breakthrough results.

4.3.1 Molecule Collision and Energy Distribution in Microwave Fields

The rapid development of communication technology and the stronger electromagnetic radiation in the environment, the study of non-thermal effects of weak microwave has once again become the focus of people's attention. In the presence of an applied electric field, polar molecules are oriented, and will align with the field direction. This could result the change of spatial collisions and the energy distribution of molecules accordingly. There is no doubt that the thermal motion makes the molecular collisions isotropic, but the polarization of polar molecules in weak microwave fields through the permittivity measurement is an incontestable fact. In this chapter, we attempt to further study the spatial collision probability and energy distribution of water molecules in the weak microwave fields by the molecular dynamics simulations [6].

GROMACS simulation package version 5.1.4 was utilized to perform all simulations. The simulation involved 500 water molecules placed in a simulated cubic box of the length of 2.47 nm and periodic boundary conditions (PBC) were applied in all three dimensions. The model of water used in SPC/E. The velocity re-scaling thermostat was utilized to maintain the equilibrium temperature at 283 K in the simulations. A cut-off distance of 1.2 nm was applied to the real-space Ewald interactions and van der Waals interactions. The particle mesh Ewald method was used to compute long-range electrostatics within a relative tolerance of 1.0×10^{-6}. The LINCS algorithm was applied to constrain the bond lengths of the hydrogen atoms. A series of equilibrium MD simulations were carried out for 10 ns in the canonical (NVT) ensemble. A leapfrog algorithm was accompanied by a time step of 1 fs and the generated trajectories stored every 5 fs.

We focus on the spatial collision of positive and negative electrodes between water molecules, that is, the collision of oxygen and hydrogen atoms. The collision is assessed by the relative displacement of collision pairs. Figure 4.15a defined the relative displacement $R(r,\theta,\varphi)$ of collision pairs, where r represents the radial distance, θ represents the zenith angle and φ represents the azimuth angle.

Figure 4.15b, c present the spatial collision probability of water molecules without and with microwave, respectively. The spatial collision area of hydrogen atoms is the shape of the paraglider. The result shows that the spatial collision of oxygen and hydrogen atoms between water molecules is not isotropic. Figure 4.16a displays the total collision number is almost unchanged by the weak microwave. However, compared to no microwave fields, the spatial collision probability of water

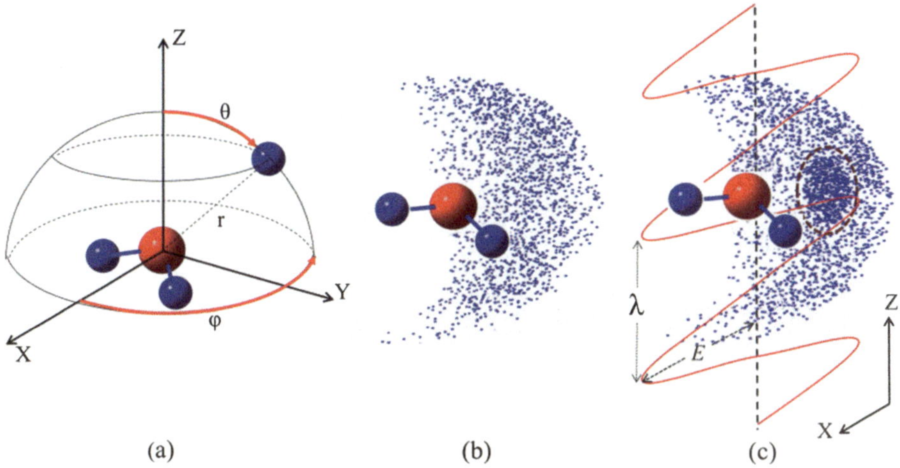

Fig. 4.15 Schematic diagram of spatial collisions of water molecules. (**a**) Definition of spherical coordinates R(r, θ, φ) for calculating spatial collision probability. (**b**) Spatial collision without microwaves. (b) Spatial collision position with microwaves

molecules increases by 3% to 9% within the range of θ from 75° to 105° and the range of φ from 165° to 195° with the weak microwave in Fig. 4.16b.

Figure 4.17 shows that the average energy is hardly changed by the weak microwave. However, we find that the weak microwave make the change of the energy distribution through carefully comparing the energy distribution with and without the weak microwave. Figure 4.18 presents that the energy distribution curves with weak microwave become higher and narrower. The illustrations in these pictures describe the proportion of medium energy molecules increases while the proportion of high energy molecules decreases. Furthermore, we also analyze the energy distribution using the extreme value function as following

$$f = A \exp\left(-\exp\left(-(e - e_c)/\delta\right) - (e - e_c)/\delta + 1\right) \tag{4.15}$$

where A, e_c, δ and are distribution amplitude, center, and width respectively. Table 4.1 lists the parameters of the energy distribution. Compared to no microwave fields, the distribution center is not changed, the distribution amplitude becomes higher, and the distribution width becomes narrower with the weak microwave. Then, we specifically analyze the distribution of high energy molecules and calculate the proportion of molecules of energy greater than 3×10^{-20} J. Figure 4.19 depicts that the proportion of high energy molecules reduced by up to 6% with the weak microwaves.

The results show that the polarization of microwave makes the change of spatial collision and energy distribution. The spatial collision probability of water molecules increases obviously in the direction of electric field and the proportion of high energy molecules decreased while the proportion of intermediate energy molecules

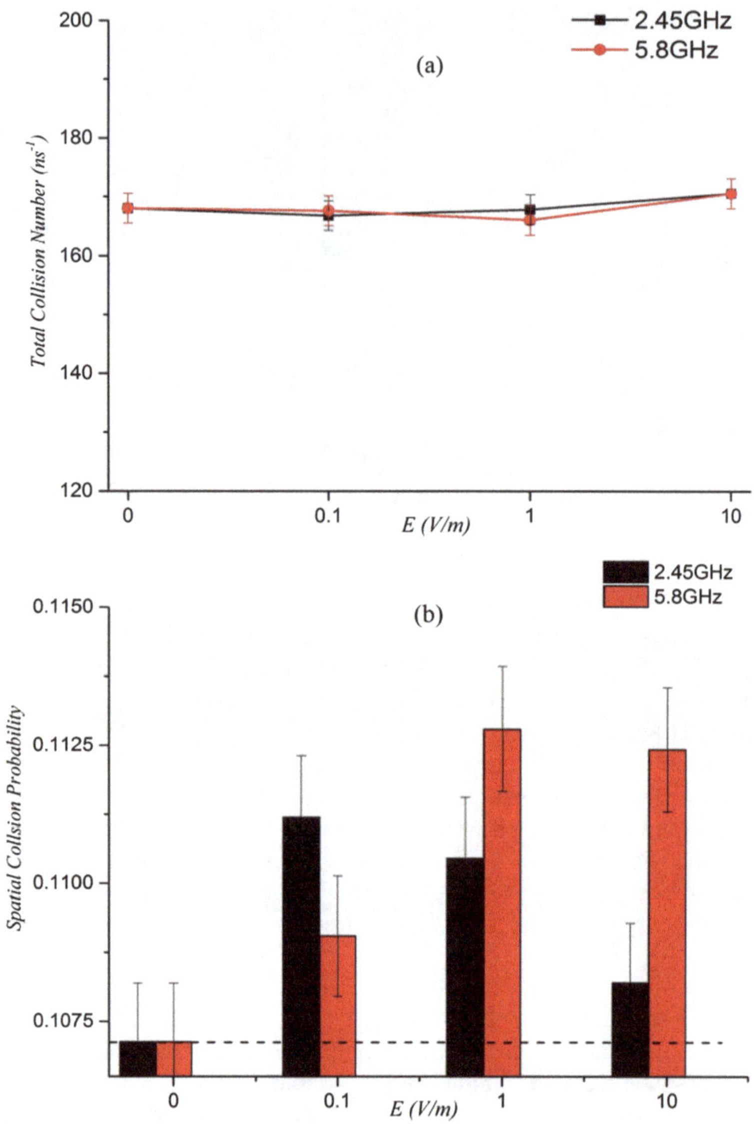

Fig. 4.16 (**a**) The total collision number of water molecules with and without the weak microwave. (**b**) The spatial collision probability within the range of θ from 75° to 105° and the range of φ from 165° to 195° with and without the weak microwave. The black horizontal dashed line represents the value without the weak microwave

increases. This shows the constrains of microwave fields on the water molecules, and these consequences would be helpful to understanding the microwave non-thermal effect and the mechanism of microwave altering activation energy and pre-exponential factor of chemical reactions.

Fig. 4.17 The average energy with and without the weak microwave

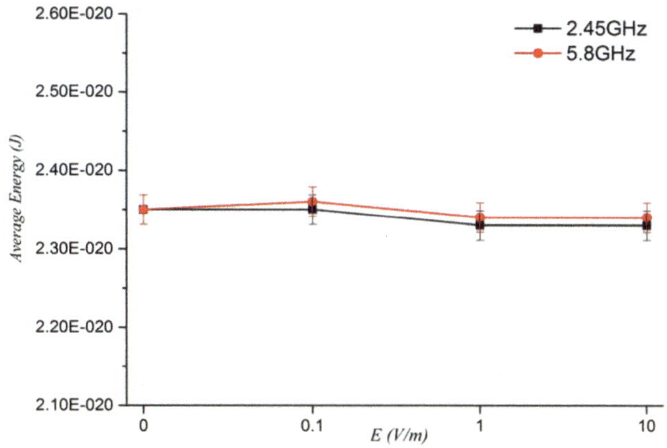

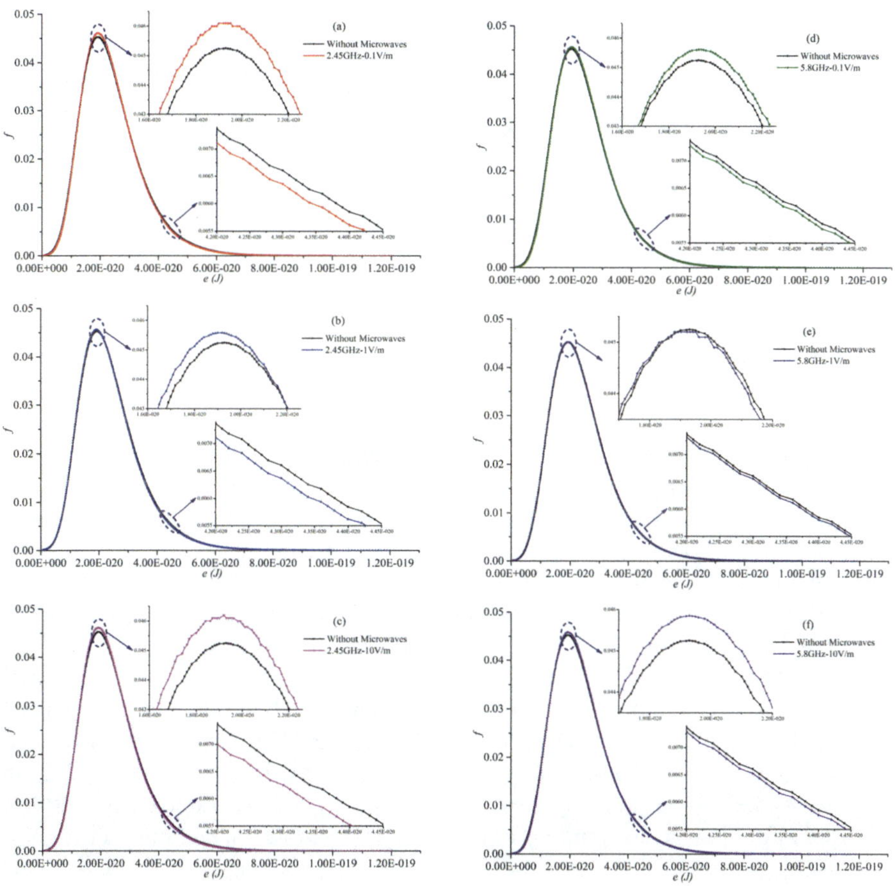

Fig. 4.18 The energy distribution without and with the weak microwave of different intensities and frequencies

Table 4.1 Parameters of the energy distribution. Distribution amplitude A (10^{-2}), distribution center e_c (10^{-20} J), and distribution width δ (10^{-21} J)

		2.45 GHz			5.8 GHz		
	No-Microwave	0.1 V/m	1 V/m	10 V/m	0.1 V/m	1 V/m	10 V/m
e_c	1.93	1.93	1.91	1.92	1.93	1.92	1.93
δ	8.26	8.09	8.19	8.08	8.18	8.28	8.17
A	4.53	4.61	4.56	4.62	4.56	4.52	4.59

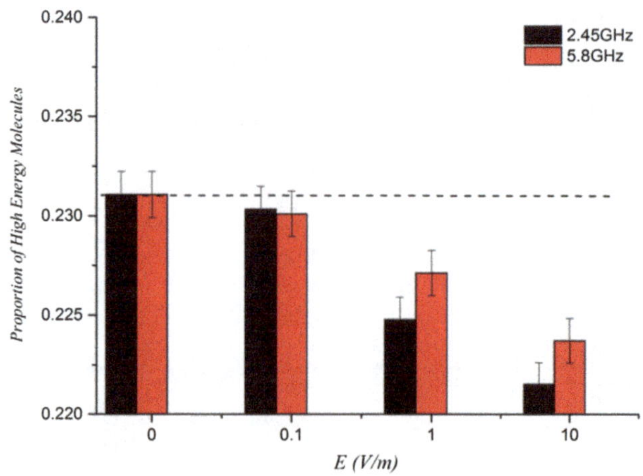

Fig. 4.19 The proportion of high energy molecules with and without the weak microwave. The black horizontal dashed line represents the value without the weak microwave

4.3.2 Desolvation Effect in Microwave Field

Solvation is the process of attraction and association of molecules of a solvent (sometime are polar solvents as their molecular structure that contains dipoles) with molecules or ions of a solute. Solvation effect is an interaction of a solute (ions) with the polar solvent (water), as polar solvent molecules can orient the appropriate partially charged portion of the molecule towards the ion in response to electrostatic attraction. This stabilizes the system and creates a solvation shell (or hydration shell in the case of water). The solvation effect has been studied experimentally for a longtime, and they play an important role in solution chemistry, biochemistry, and some aspects of cell and membrane biology.

The molecular dynamics simulations for investigating the effect on properties of NaCl aqueous solutions were performed with the free simulation package GROMACS 5.1.2. The water model is OPC model. The simulations involved 0.1 M, 0.5 M and 1.0 M (molality), corresponding to 5–50 ion pairs and 2896 water molecules placed in a cubic periodic box with periodic boundary conditions, using minimum image convention. Long-range Coulombic interactions were evaluated using a cutoff of 1.0 nm, the short-range interaction cutoff was set also to1.0 nm. A leap-frog algorithm was used with a time step of 1 fs. All systems

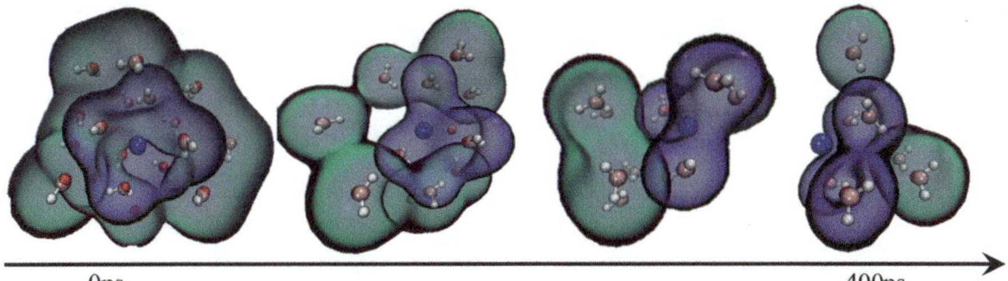

Fig. 4.20 Snapshots of the hydration shell of sodium ions in the microwave field

were first simulated in the NVT ensembles for at least 2 ns to establish thermal equilibrium, and at least 4 ns in the NPT ensembles to establish pressure balance, pressure was controlled by the Parrinello-Rahman thermostat barostat employing pressure damping factors of 2.0 ps. In order to isolate the field effects from the thermal effects, the microwave fields were applied in the NVT ensemble, then, the series production simulations were nonequilibrium NVT simulations, the ionic solutions were simulated for 4000 ps and the generated trajectories stored every 0.05 ps. In these simulations, temperature was controlled by the Nose-Hoover thermostat using a temperature damping factor of 0.8 ps.

Our simulations monitor the effects of changing microwave field strength, concentration, hydration shell size. In Fig. 4.20, we plot the hydration shell size of sodium ions versus the time elapsed after applying the microwave field ($E = 5 \times 10^9$ V/m). A water molecule is considered to be part of the shells as long as it remains within 0.45 nm of any sodium ions, the hydration shell size analysis is performed every 50 fs.

As the microwave field accelerates the shells, after some time, them begin to collide with other water molecules, in Fig. 4.20, there is an immediate loss of water molecules, we find roughly 49% of water molecules leave the shells within 400 ps, but the last 5–6 water molecules remain in the shell of every sodium ions. In the case of the microwave field, the field oscillates back and forth, leading to the induced velocities of shells changing direction in response to the field after a frictional lag time, showing us a cyclical behavior in the shell size curve.

In order to find out what will happen to hydrogen bonds (HBs) when the hydration shells are weakened by the microwave fields ($1 \times 10^4 < E < 6 \times 10^4$ V/m), a HB analysis is performed. The HB lifetime τ_{HB} is given by the inverse forward rate constant k estimated by work of Spoel et al., $\tau_{HB} = {}^{1}\!/_{k}$, HB autocorrelation as a function of time for water molecules in the shell was computed by model of Luzar and Chandler. τ_{HB} in shells of Na^+ at different concentrations under the effect of microwave field are listed in Table 4.2. As shown in Table 4.3, τ_{HB} and decay times of autocorrelation function of 0.1 M and 1.0 M NaCl solution are subtly weakened by the weak the microwave fields. The result might be explained by the formation of hydrogen bonds within the second hydration

Table 4.2 τ_{HB} for HBs in the hydration shells of ions

Concentration (mol/L)	τ_{HB} for HB in shells of Na^+(ps)			
	$E = 0$ V/m	$E = 1 \times 10^4$ V/m	$E = 3 \times 10^4$ V/m	$E = 6 \times 10^4$ V/m
0.1	4.52	4.21	4.69	4.73
0.5	4.62	4.64	4.60	4.58
1.0	4.71	4.60	4.67	4.71

Table 4.3 Calculated and measured value of σ without microwave field

	0.1 M	0.5 M	1.0 M
Measured σ (s/m)	0.98	4.78	9.1
Calculated σ (s/m)	0.73	4.21	7.41

shell and by water molecules exchanging between second and first hydration shell. When the microwave field is present, shells velocity is low, collisions are not strong enough to remove water molecules, the affection is subtle and unstable. The collision induced by the microwave field could just disturb the outer layer of hydration shells, leading water molecules more likely to loss and easier to exchange, the life of HBs are slightly affected then.

The electrical conductivity (σ) can be estimated using:

$$\sigma = \frac{F^2|A|^2 C(D_+ + D_-)}{RT} \tag{4.16}$$

Where A is the valence, F is Faraday constant, R is the ideal gas constant, T is the absolute temperature, C is the concentration and D is the diffusion coefficients of cation or anion. The results are shown in Table 4.3. As mentioned previously, these results are caused by the microwave fields, with the fields shattered (or weakened) the hydration shells of ions, less water molecules exist in the hydration shells, HBs more likely to broken, and stronger microwave fields were shown to lead to freer ions, Na^+ and Cl^- easier to drift under the effect of strong microwave field. Our values for the Δσ ($\Delta\sigma = (\sigma E = 1 \times 10^4 - \sigma E = 0)/\sigma E = 0$) of 0.1 M NaCl solutions at the microwave field of $E = 1 \times 10^4$ V/m is 6.1%, the Δσ of 0.5 M NaCl solutions at the microwave field of $E = 1 \times 10^4$ V/m is 5.2% and Δσ of 1.0 M NaCl solutions is 8.5%. The results are similar to our former experiment results of 2.1% and 4.0% for 0.1 M and 0.5 M NaCl solutions, respectively.

4.3.3 MD Simulation of Special Permittivity of Mixed Solution

Herein, we have carried out molecular dynamics simulation for different mixed systems, namely, N,N-dimethylacetamide (DMAC) – water mixtures and Dimethyl Sulfoxide(DMSO) – water mixtures, to observe the special permittivity phenomenon under microwave field. These studies explained the special dielectric properties of the mixtures, which are in good agreement with the experimental results.

4.3.3.1 N,N-Dimethylacetamide (DMAC) − Water Mixtures

The simulations were performed using a modified version of the GROMACS 4.5.5 [7] simulation package with the TIP4P [8] model and the OPLS-AA potential characterizing the water and DMAC molecules, respectively. The simulations involved 555 water molecules placed in a cubic simulation periodic box and two concentrations of DMAC, i.e., 0.8 M and 8 M (molality), corresponding to 8 DMAC molecules and 80 DMAC molecules, respectively. TIP4P water molecules were added to the system from a relaxed liquid configuration at 298 K and 1 bar.

Prior to the system relaxation MD, energy minimization was carried out using a composite protocol of the steepest descent, conjugate gradient, and truncated Newton steps, using termination gradients of 5000, 1000, and 100 kJ/mol/nm, respectively. MD was coupled to a modified Berendsen-thermostat [9] system. The particle mesh-Ewald method [9, 10] was used to compute long-range electrostatics to within a relative tolerance of 1×10^{-6}. A cut-off distance of 8 Å was applied to real-space Ewald interactions. The same value was used for Van der Waals interactions. The LINCS algorithm [11] was applied to constrain the bond lengths to hydrogen atoms and to maintain the rigid TIP4P geometry. A leap-frog algorithm [12] with a time step of 1 fs was used. The criterion for the hydrogen bonds is defined [13] as $r \leq r_{HB} = 0.35$ nm, $\alpha \leq \alpha_{HB} = 30^0$, the value of $r_{HB} = 0.35$ nm corresponds to the first minimum of the RDF of SPC water.

Here, a smaller subset of the radial distribution function (RDF) was studied to detect any structural change in the DMAC aqueous solutions under the E/M field. Figure 4.21 presents the RDFs of H–H, O–O, H–O, N–N, N–O, and N–H in the DMAC 0.8 M aqueous solution both with and without an E/M field. Figure 4.21a–c shows that the height of the first peak in each set of profiles decreases as the intensity of the E/M field increases. However, Fig. 4.21d–f shows that the height of the first peak in each set of profiles increases with the intensity of the E/M field. The first minimum decreases with increasing field strength, as seen in Fig. 4.21a, b. This result indicates that the structure of the first hydration shell of DMAC changes significantly with the application of an E/M field.

Figure 4.22 presents the RDFs of H–H, O–O, H–O, N–N, N–O, and H–O in an 8 M DMAC aqueous solution with and without an E/M field. From the figure, a more obvious variation in the height of the first and second peaks of the DMAC solution is observed than that for the 0.8 M solution, especially when the E/M field strength reaches a maximum. It is also found that the positions of the first peaks of $g_{OO}(r)$, $g_{HO}(r)$, and $g_{HH}(r)$ almost do not shift. However, the positions of the second peak of $g_{HO}(r)$ shift more than 0.02 Å. These results indicate that the structures of both the first and second hydration shells of DMAC change significantly with the application of an E/M field.

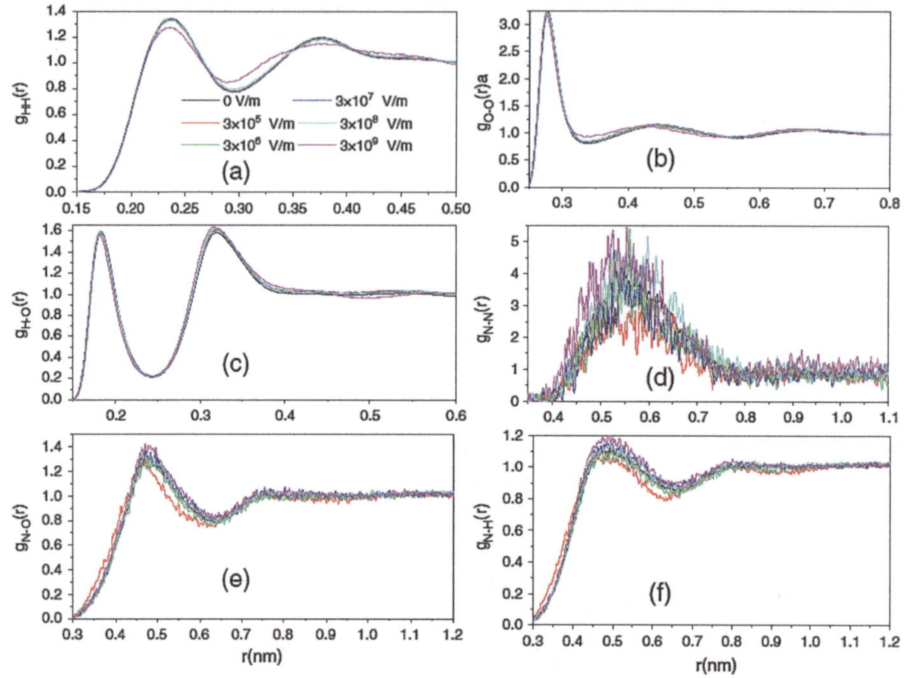

Fig. 4.21 RDF of 0.8 M DMAC aqueous solution with and without an applied E/M field. (**a**) H–H, (**b**) O–O, (**c**) H–O, (**d**) N–N, (**e**) N–O, and (**f**) N–H

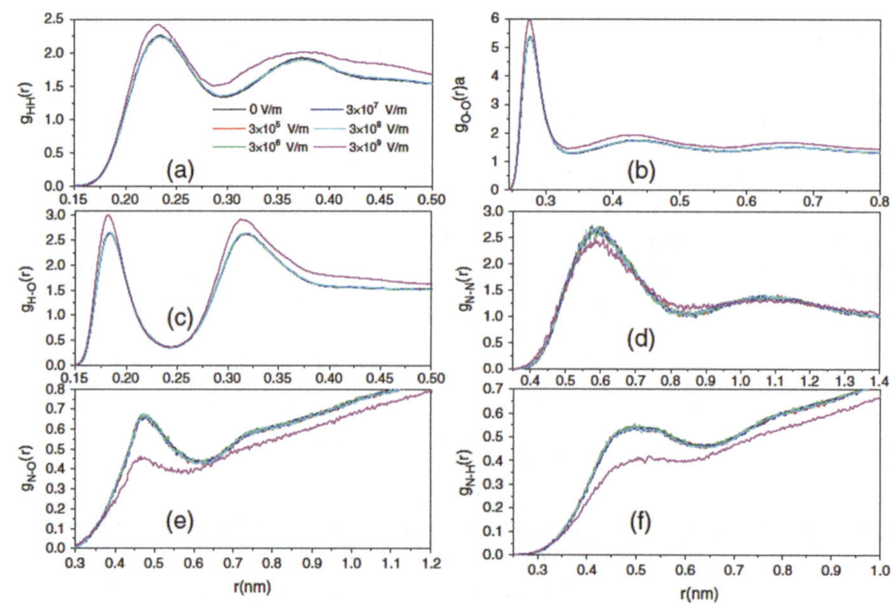

Fig. 4.22 RDF of 8 M DMAC aqueous solution with and without the application of E/M field. (**a**) H–H, (**b**) O–O, (**c**) H–O, (**d**) N–N, (**e**) N–O, and (**f**) N–H

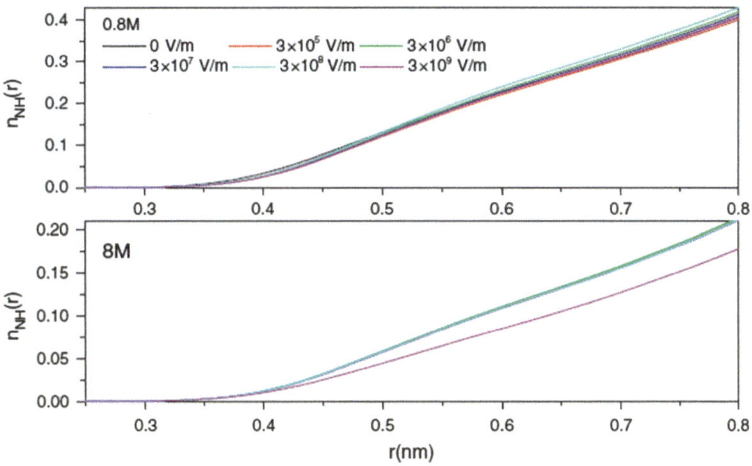

Fig. 4.23 Running coordination numbers of N atoms as a function of distance r (nm) for H atoms under E/M field; H represents the hydrogen atoms of water and N represents the nitrogen atoms of DMAC

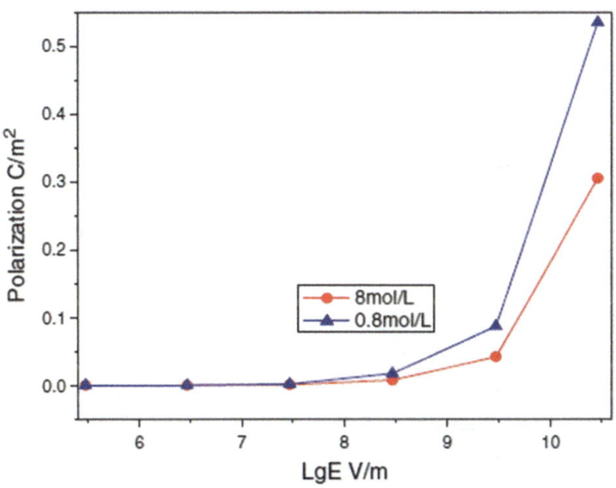

Fig. 4.24 Molecular polarizability of the DMAC aqueous solutions under different E/M field

Figure 4.23 indicates the running coordination numbers of the nitrogen atom as a function of distance *r* (nm). Due to the less availability of water molecules, the running coordination numbers for the 8 M solution decrease remarkably compared to the 0.8 M solution. This reveals the randomization of the position and orientation of water molecules, and thus, the weakening of the water network. The water shell that contributes to the complex permittivity of the system consists of a tiny unit of

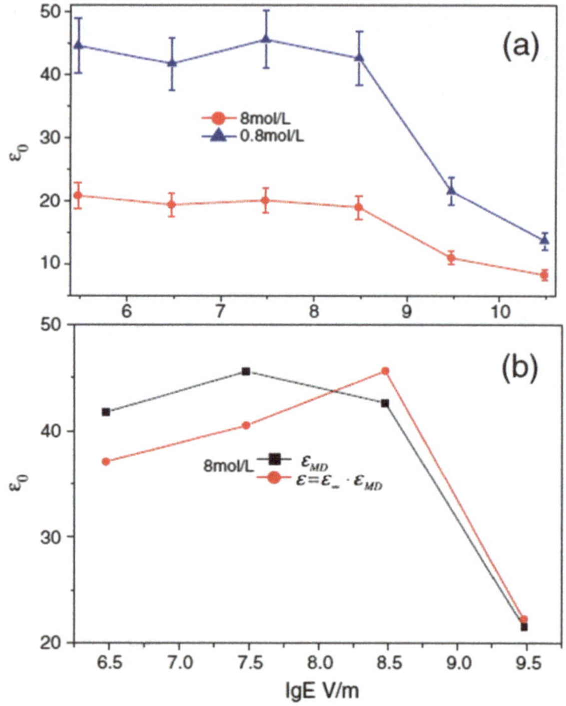

Fig. 4.25 Static complex permittivity of DMAC aqueous solutions as a function of E/M field

hydrogen bound to the water molecules, which results in a decrease in the effective number of aligned dipoles. Further confirmation can be obtained from Fig. 4.24.

Figure 4.24 illustrates the variation in the molecular polarizability of the DMAC aqueous solutions with concentrations of 0.8 M and 8 M, as a function of the E/M field. As can be seen from this figure, the molecular polarizability gradually increases with the E/M field strength, which shows that the direction of the molecular dipole moment is gradually aligned in the same direction as that of the E/M field.

The static complex permittivity of the DMAC aqueous solutions at 298 K under various E/M fields is obtained (Fig. 4.25). As shown in Fig. 4.25a, there seems to be a threshold effect at about 3×10^8 V/m. When the electric field strength is above 3×10^8 V/m, the complex permittivity declines with increasing E/M field strength for both the 0.8 M and 8 M solutions. When the electric field strength is below 3×10^8 V/m, the static complex permittivity changes slightly. Figure 4.25b shows the complex permittivity, ε_{MD}, and the total complex permittivity, ε; the error is between 3% and 12%. This indicates the contribution of the non-polarizable force fields TIP4P and OPLS-AA to the complex permittivity.

Figure 4.26 shows that the application of an E/M field has a major influence on the average hydrogen-bond number. It is found that for 0.8 M, the average hydrogen-bond number tends to increase with the E/M field strength, but for 8 M, it tends to decrease with the E/M field strength. The average hydrogen-bond numbers for 0.8 M

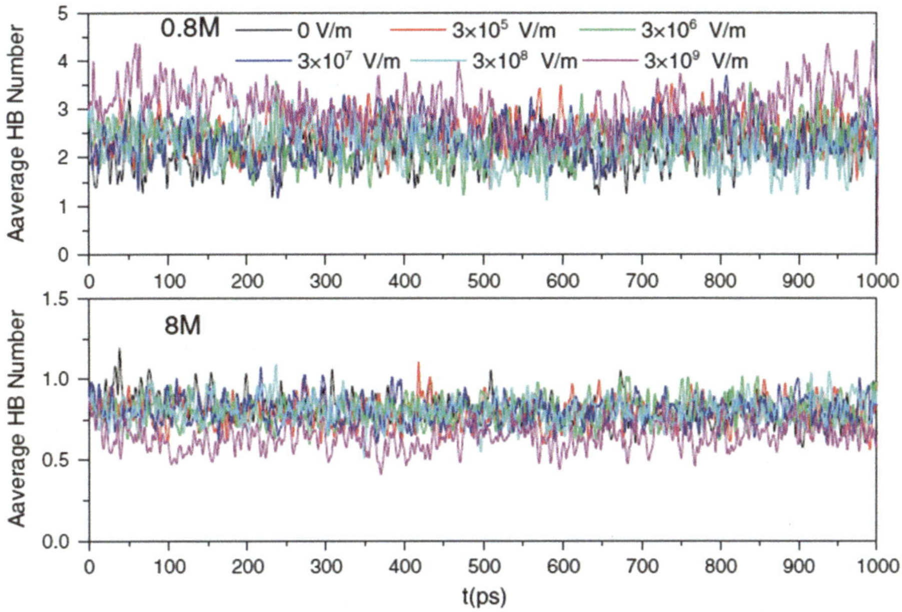

Fig. 4.26 Average hydrogen-bond number as a function of time for DMAC–water for both the 0.8 M and 8 M solutions under the E/M field

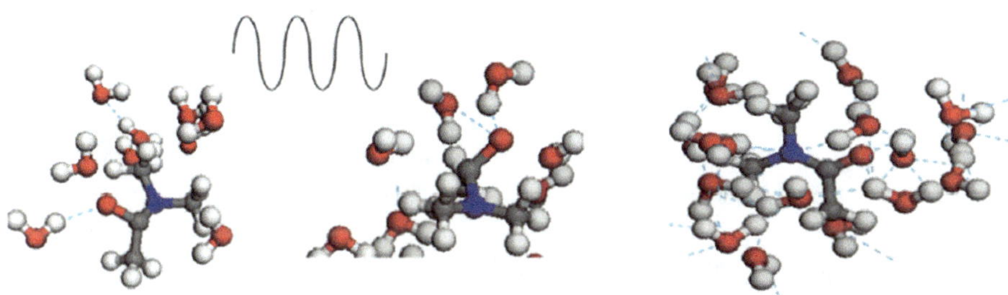

Fig. 4.27 Kinetics of the hydrogen-bonds between DMAC–water under an E/M field

are larger than 8 M. This is attributed to the fact that the E/M field weakens the DMAC–water interactions at 8 M, which orientates some of the water dipoles in the opposite direction, leading to a decrease in the dipole contribution to the average hydrogen-bond number, and vice versa for the 0.8 M solution (Fig. 4.27).

From Fig. 4.28, the hydration-bond lifetime, τ_H, is estimated from the long-time decay of the autocorrelation functions of CHB(t) by an exponential CHB(t) = exp. $(-t/\tau_H)$, same as the method used in other reports [14–17]. Note that the goal here is not to obtain an accurate value but just to find a trend as a function of the E/M field strength. It is found that for both the 0.8 M and 8 M solutions, the hydrogen-bond lifetimes decrease as the E/M field strength increases. At 0.8 M, the E/M field

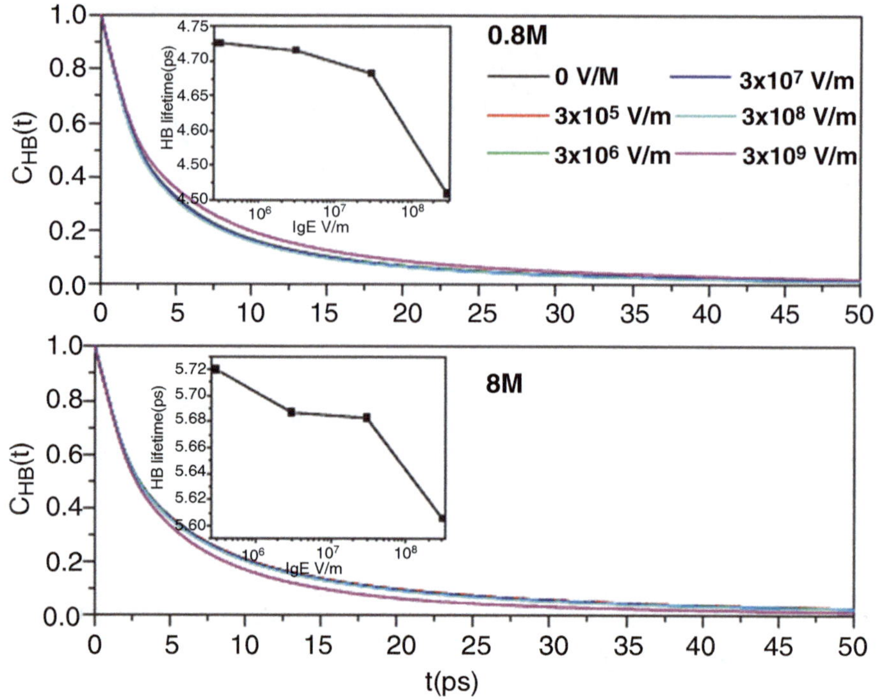

Fig. 4.28 Hydrogen-bond autocorrelation as a function of time for DMAC–water at 0.8 M and 8 M under the effect of E/M field. Inset in Fig. 4.14 shows the HB lifetime as a function of the E/M field

enhances the DMAC–water interactions, which results in the same trend of RDF for N–H (Fig. 4.21f) and O1–H (Fig. 4.21a) with hydrogen-bond lifetime. As shown in Figs. 4.21e, f and 4.29a, the hydration ability of DMAC is enhanced by increasing the E/M field, which leads to the hydrogen-bond lifetime increasing with the E/M field strength. For 8 M, the E/M field enhances the water–water interactions, leading to the same trend as that in the RDF for N–H (Fig. 4.22f) and O1–H (Fig. 4.29b) with hydrogen-bond lifetime. As shown in Figs. 4.22e, f and 4.29b, the hydration ability of DMAC is weakened by increasing the E/M field. When the intensity of the external E/M field is up to 3×10^9 V/m, the dipole energy ($E_{OP0} \approx 4.8$ eV) is larger than the hydrogen-bond energy of 2.5 eV per bond, leading to fluctuations in the hydrogen-bond lifetimes for both systems.

4.3.3.2 Dimethyl Sulfoxide (DMSO) − Water Mixtures

The dielectric properties of water−DMSO mixtures of different mole ratios have been investigated in the range 1 GHz to 40 GHz at 298 K by using MD simulations. The mixtures with DMSO mole fractions xDMSO (xD) = 0.1, 0.21, 0.35, 0.48, and

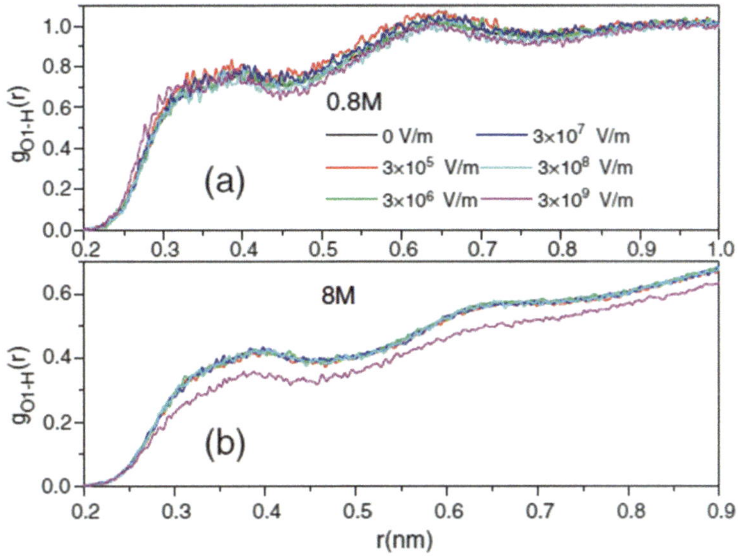

Fig. 4.29 Radial distribution functions of 0.8 M and 8 M DMAC aqueous solutions with and without an applied E/M field. O1 represents the oxygen of DMAC and H represents the hydrogen of water

0.91 and the intermolecular potentials are described by site−site interactions given by the sum of the Lennard−Jones plus Coulomb terms:

$$u_{ij}(r) = 4E_{ij}\left[\left(\frac{d_{ij}}{r_m}\right)^{12} - \left(\frac{d_{ij}}{r_m}\right)^{6}\right] + \frac{q_i q_j}{4\pi\varepsilon_0 r_m} \quad (4.17)$$

where r_m is the distance between the *i*th molecule and the *j*th molecule in the system, q_i and q_j are the effective charges of sites i and j, respectively, and E_{ij} and d_{ij} are the energy and distance parameters in the Lennard-Jones potential. They are determined by the Lennard-Jones parameters of the different sites according to the "mixing" rule:

$$d_{ij} = (d_i + d_j)/2, E_{ij} = \sqrt{E_i E_j} \quad (4.18)$$

where q_i, d_i, and E_i indicate the oxygen, sulfur, and methyl sites in DMSO and the oxygen and hydrogen sites in water, as shown in Table 4.4.

The simulations reported in this study were run in the NVT ensemble on systems comprising a thousand molecules placed in a cubic box with periodic boundary conditions at an average temperature of 298 K. The box length was chosen to match the experimental density of the DMSO−water mixture solvent at 298 K. The Lennard-Jones forces were cut off to half of the box length and long-range Coulomb

Table 4.4 L-J parameters and partial charges for water and DMSO

	$E/(\text{kJ mol}^{-1})$	$d/(\text{nm})$	$q/(a.\ u.)$
O (water)	0.65	0.3165	−0.82
H (water)	0.0	0.0	0.41
O (DMSO)	0.30	0.28	−0.459
S	1.00	0.34	0.139
CH_3	1.23	0.38	0.16

Table 4.5 Composition of the DMSO-water mixtures. Solvent 1 is DMSO; solvent 2 is water

Composition	x_1	x_2	n_1	n_2	L(Å)	ρ^{a} (g/cm^3)	Refractive index (n_D)
1	0.1	0.9	100	900	34.05	1.0105[b]	1.3458[b]
2	0.21	0.79	210	790	36.76	1.0242[b]	1.3600[b]
3	0.35	0.65	350	650	39.0	1.0927	1.4385
4	0.48	0.52	480	520	41.38	1.0983	1.4525
5	0.91	0.09	910	90	47.93	1.0965	1.4721

interactions were treated using the Ewald sum method. The equations of motion were solved using the combined SHAKE and leap-frog algorithms with a time step of 1 fs, during uninterrupted production runs, which lasted for about 50 ps for each mixture. Then, the simulations were run in the NVE ensemble, and all simulations were extended up to 1.2 ns, where the first 50 ps were considered as equilibration. The starting solvent configurations for the five mixtures were chosen according to the specifications given in Table 4.5.

The composition dependence of the calculated values of ΔG is given in Fig. 4.30. The figure shows that the free energy of activation increases with the concentration of the nonelectrolyte in water, reaching a maximum value of 3.5 at a concentration fraction of 0.35. This difference can be attributed to the fact that the components are capable of specific interactions with each other.

The theory of rate processes to molecular dipole relaxation was applied with the following equation

$$\tau = (\mathrm{h}/\mathrm{K_B T}) \exp{(\Delta \mathrm{G}/\mathrm{RT})} \tag{4.19}$$

where ΔG is the free energy change of activation for the dipole relaxation, K_B is Boltzmann constant, h is Planck constant and R is the molar gas constant.

The complex permittivity are fitted by the nonlinear least-squares fit method to the Havriliak-Negami expression to obtain various dielectric parameters:

$$\varepsilon(\omega) = \varepsilon_\infty + (\varepsilon_0 + \varepsilon_\infty)/(1 + j\omega\tau) \tag{4.20}$$

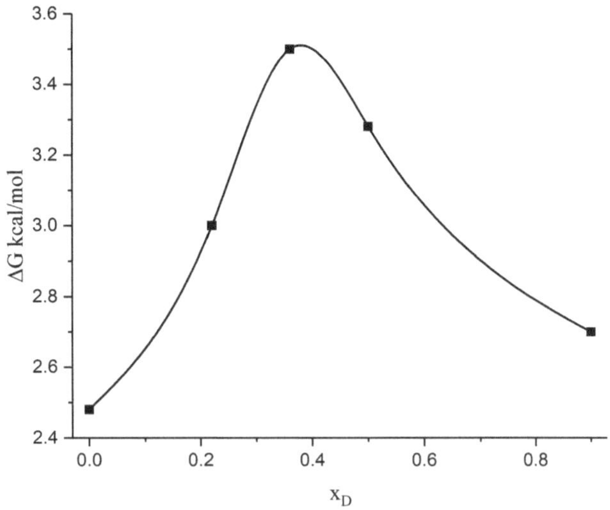

Fig. 4.30 Free energy change of activation of DMSO–water solutions vs. mole fractions

where $\varepsilon(\omega)$ is the complex permittivity at an angular frequency ω, ε_∞ is the permittivity at high frequency, ε_0 is the static dielectric constant.

Permittivity and losses for DMSO–water mixtures at 298 K are shown in Fig. 4.31b, where a single relaxation peak is observed for the entire concentration range of all DMSO–water mixtures in the frequency range of 1–40 GHz. Figure 4.31a clearly indicates that the higher the frequency, the lower is the real part of the permittivity (ε'). Nevertheless, the real part of the permittivity (ε') decreases with increasing concentration of DMSO; however, a completely opposite result is obtained when ε' is more than 0.35. It can be deduced that the ε' of the mixtures strongly depends on the DMSO concentration, which is attributed to the cooperative motion of DMSO–water molecules through hydrogen bonds. Figure 4.32 shows the dependence of the imaginary part of permittivity on the mole concentration of DMSO at the microwave frequency of 2.45 GHz. The calculated results are consistent with the results of the special complex permittivity of binary mixtures in the experimental measurement.

The static complex permittivity, the dielectric relaxation at the frequency of microwaves, as well as the microwave spectra of DMSO–water mixtures, have been investigated through MD simulations at room conditions, and these simulations have proven to be an efficient tool for the study of molecular processes in solutions over a sufficiently wide frequency range.

This method was also used to calculate the composition-dependent dielectric properties of the DMSO–water mixtures. The static complex permittivity, the dielectric relaxation in the frequency range of microwaves, as well as the microwave spectra of DMSO–water mixtures, over the whole composition range at room conditions have been investigated by MD simulations. There is a good agreement

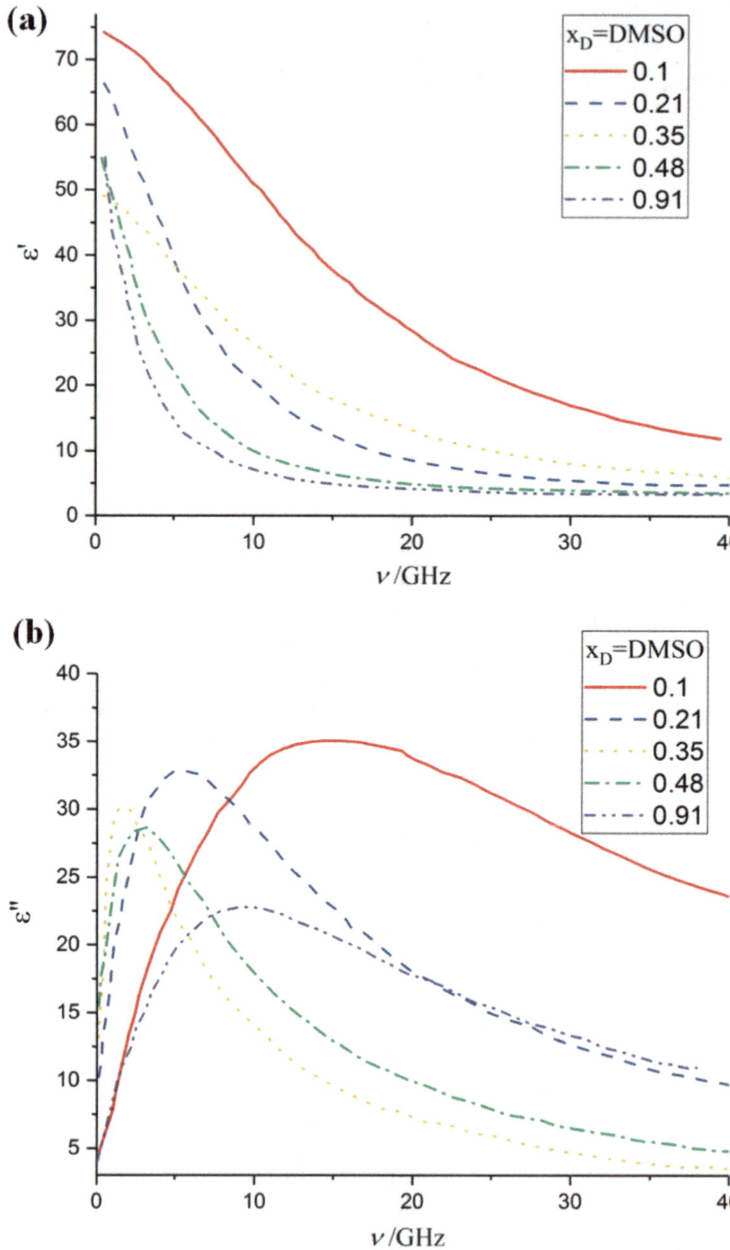

Fig. 4.31 Permittivity and losses for DMSO–water mixtures at various concentrations at 298 K. (**a**) The real part of the permittivity vs. frequency. (**b**) The imaginary part of the permittivity vs. frequency

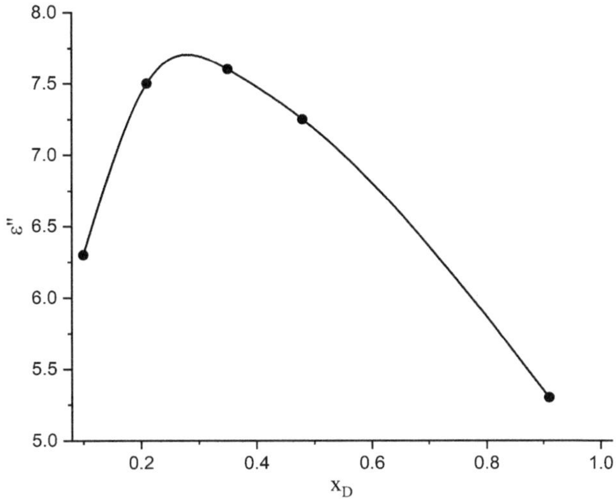

Fig. 4.32 Mole fraction dependence of the losses of DMSO-water mixtures at 2.45 GHz

between the simulated and available experimental complex permittivity for all solution compositions. Although the intermolecular potentials used here are not fully polarizable and no adjustment of the parameters has been made, the simulated values for ε_0 are only about 10% lower compared to the experimental ones. This behavior is, in turn, consistent with the formation of H-bonded aggregates of the type DMF: $2H_2O$, as well as with the experimental observations. The mixture's overall dielectric relaxation time τ_D is in good agreement with the experimental principal dielectric relaxation time.

4.3.4 Density Functional Theory (DFT) Simulation of Nonlinear Dielectric Properties with Microwaves

Usually, the dielectric properties represent the interactions between microwave and materials. The measurement of dielectric properties of the polar solutions is an useful method to study the mechanism between molecular clusters and binary molecular clusters. Aiming at inquiry the mechanisms, our group performed two experiments to measure the dielectric properties of DMSO-ethanol mixtures. Surprisingly, we found an abnormal phenomenon that the imaginary part ε_r'' of the mixtures relative permittivity is larger than those of their pure substances at some concentrations, as shown in Fig. 4.33. Besides, to further explore the unusual behavior of the relative permittivity, low-intensity microwaves were applied to the DMSO–ethanol mixtures

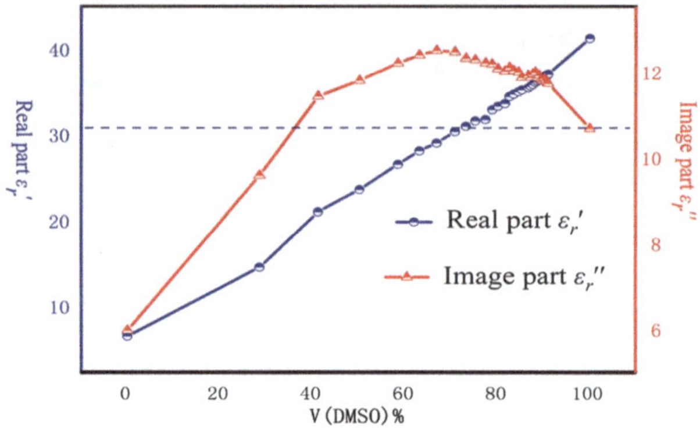

Fig.4.33 The relative permittivity of DMSO-ethanol mixtures at 2.45 GHz and at 20 °C as a function of the volume fraction of pyridine V (DMSO)%. The blue and red curves represent the real part ε_r' and the imaginary part ε_r'', respectively

as shown in Fig. 4.33. In this case, the dielectric properties of the mixtures behavior nonlinearly, while those of the pure components vary linearly.

The experimental results validate the existence of a special microwave effect in low-level-intensity electric fields. Further theoretical work on the details of the interaction mechanism between the microwaves and mixtures should be studied from the molecular level. For this purpose, we put forward a conjecture that there exist some abnormal molecular clusters in mixture leading to the property changes of the DMSO–ethanol mixtures to explain the abnormal phenomenon as mentioned above.

An external electric field has been employed to investigate the structural evolution of molecular clusters $(DMSO)_m(C_2H_5OH)_n$ ($m = 0$–4; $n = 0$–4; $m + n \leq 4$). The field strengths are increased gradually from 0 to 1.028×10^9 V/m with the step of 1.028×10^8 V/m along the direction of permanent dipole moments in molecular clusters. The structure predication of the initial molecular clusters $(DMSO)_m(C_2H_5OH)_n$ are based on the ABCluster software [18]. Further calculations are performed with Gaussian 03 program package [19], Multiwfn program [20] and VMD [21]. The local structure optimizations of all molecular clusters are performed in solutions by the M06-2X approach in conjunction with the 6–311++G(d,p) basis set [22] using Gaussian 09 programs.

By calculating the hydrogen bond energies of the dimers (0.013 a.u. in $(C_2H_5OH)_2$; 0.015 a.u. in $(DMSO)(C_2H_5OH)$), we find that the hydrogen bond of $(C_2H_5OH)_2$ molecular cluster is much weaker than that of $(DMSO)(C_2H_5OH)$. In analogy with the explanation of the DMSO-water system, it is found that new molecular clusters $(DMSO)_m(C_2H_5OH)_n$ in the DMSO-ethanol mixtures is formed due to DMSO molecule breaking the hydrogen bonding interaction within ethanol clusters $(C_2H_5OH)_n$.

For the DMSO-ethanol mixtures, an interesting phenomenon, that the imaginary part ε_r'' of the mixtures relative permittivity is larger than those of their pure substances at certain concentration, was found, as shown in Fig. 4.33. To some extent, ε_r'' namely the imaginary part of relative permittivity in mixtures, represents power dissipation during the relaxation process, which is related to the properties of molecular clusters in mixtures. For example, the power dissipation increases with the size of molecular cluster. To clearly understand and interpreted the exceptional phenomenon on the dielectric spectrum of the DMSO-ethanol mixtures, a systematical investigation about the cluster radii of the molecular clusters ($(DMSO)_m(C_2H_5OH)_n$) have been listed in Figs. 4.34, 4.35 and 4.36 by using DFT calculations. Here, the cluster radius represents the farthest distance between the geometric center of molecular cluster and its molecular van der Waals (vdW) surface. It is clear from Fig. 4.34 that the cluster radii of certain binary molecular clusters are larger than pure molecular clusters, when the number of molecules in binary molecular clusters is the same as that in the pure molecular clusters. For the cases where the binary molecular clusters along with larger cluster radii keep up with the change in direction of microwave fields, more molecular friction and dielectric loss will be produced. Therefore, through the analysis between the experimental phenomenon and our calculated results, we can speculate that the existence of the binary molecular clusters in DMSO-ethanol mixtures with large sizes maybe the intrinsic reasons of the exceptional phenomenon on the imaginary part ε_r''.

Under external electric field, the root-mean-square deviation (RMSD) can be take as a decision foundation of whether changes will take place for the micro-structure of molecular clusters. For the purpose of understanding how the properties of DMSO-ethanol molecular clusters change with different electric field strength, the RMSD between the structures of molecular clusters $(DMSO)_m(C_2H_5OH)_n$ (m = 0–4; n = 0–4; $m + n \leq 4$) are also calculated. The results indicate that the RMSDs of pure molecular clusters ($(DMSO)_m$, m = 2–4; $(C_2H_5OH)_n$, n = 2–4)) only produces a small change with electric field strengths increasing from 1.028×10^8 V/m to 1.028×10^9 V/m, which is in accordance with Rai's results [23, 24]. However, to our surprise, there exist some abnormal binary molecular clusters, such as 1D3E-III, 2D1E-III and 2D2E-IV, whose RMSD changes obviously and significantly with electric field strengths increasing. In order to clearly show the structure mutation of 1D3E-III, 2D1E-III and 2D2E-IV, Figs. 4.35 and 4.36 depict their field-induced structure evolution along with the cluster radii and the hydrogen bond length. In Fig. 4.37, one can see that the structures and the cluster radii of molecular clusters has changed suddenly as the electric field strength increases to some characteristic values. In the meantime, the hydrogen bond lengths between DMSO/ethanol and ethanol molecules in molecular clusters also vary correspondingly with the structure evoluting. The electric field has different influence on the structures of the three

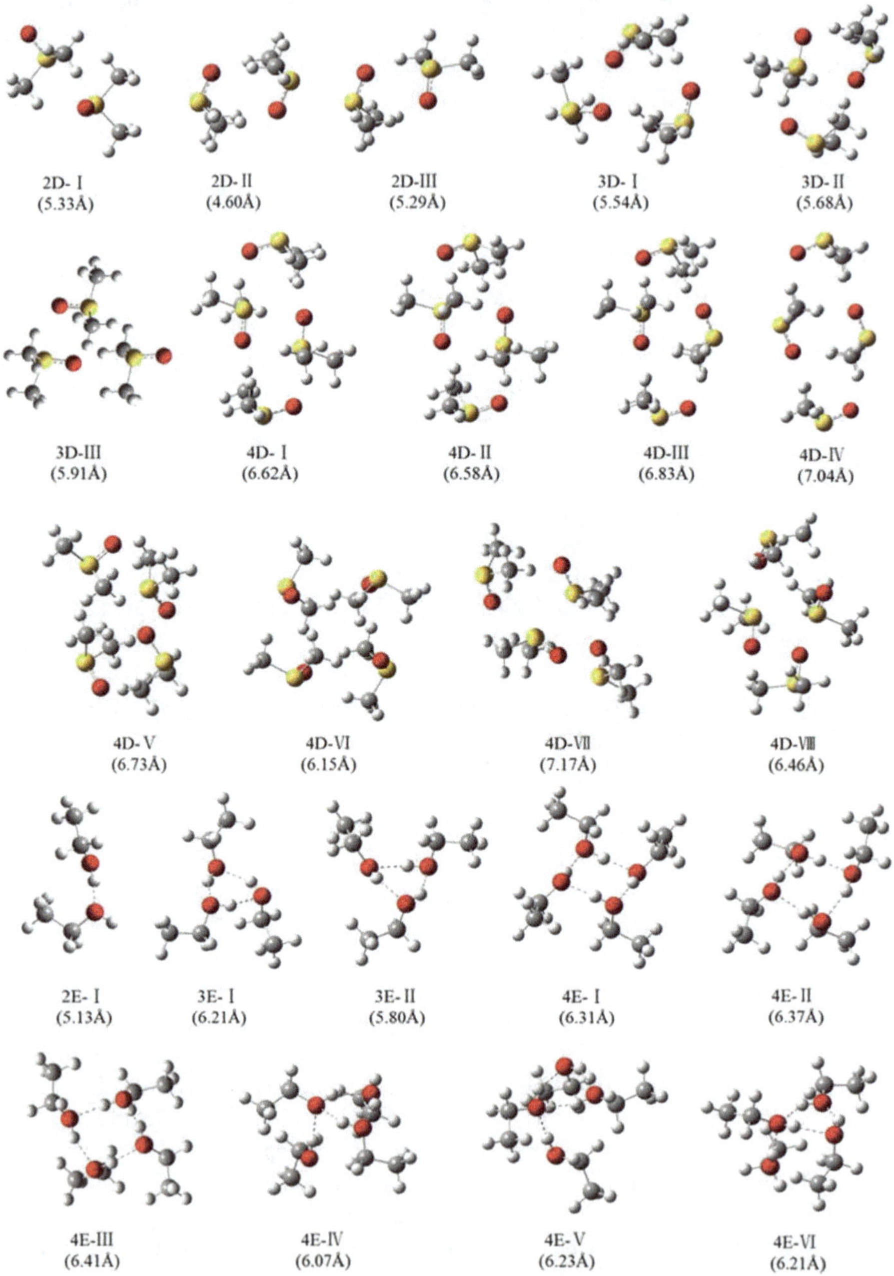

Fig. 4.34 Optimized structures of molecular clusters along with its cluster radii. The former three lines are $(DMSO)_m$ (m = 2–4), and the the latter three lines are $(C_2H_5OH)_n$ (n = 2–4). The red, yellow, gray and white balls represent oxygen, sulfur, carbon and hydrogen atoms, respectively

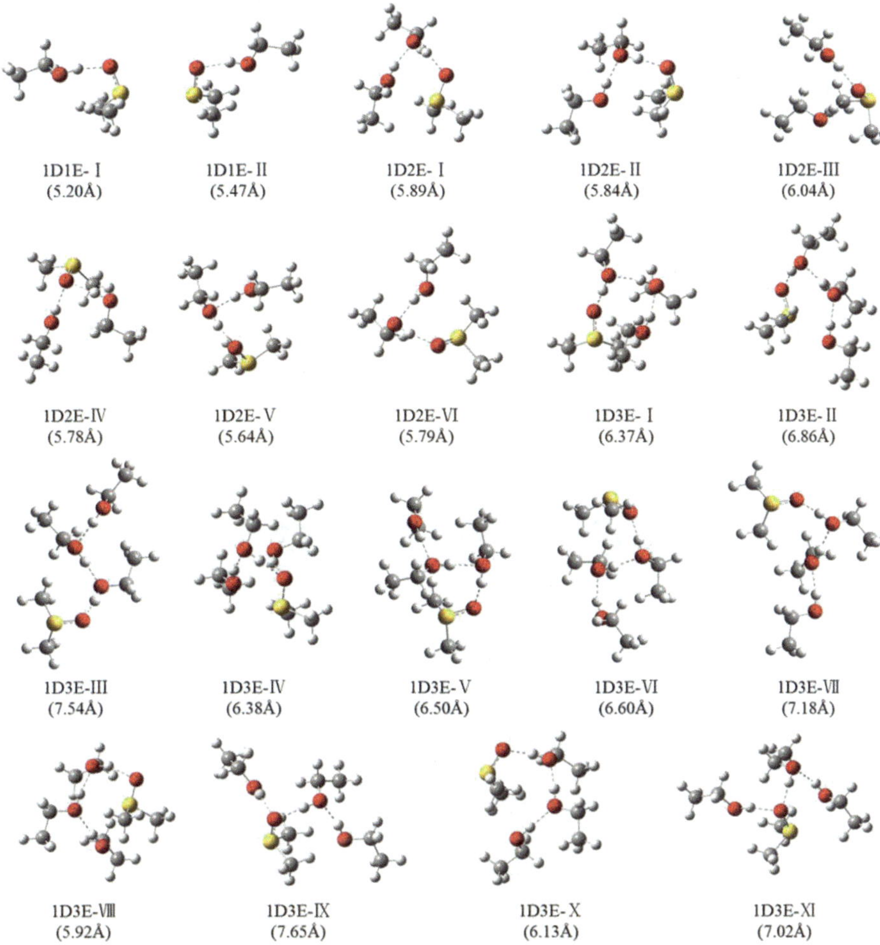

Fig. 4.35 Optimized structures of (DMSO)(C_2H_5OH)$_n$ (n = 1–3) molecular clusters along with its cluster radii

molecular clusters, and had the highest contribution to the change in the structure of 1D3E-III.

The physical properties of mixtures are closely related with its micro-structure. The micro-structure evolutions of molecular clusters existed in mixtures are embodied in the variations of the physical properties of the mixtures. Therefore, our calculated results validates that the existence of some binary molecular clusters can provide a reasonable explanation for the abnormality of the dielectric properties of mixtures.

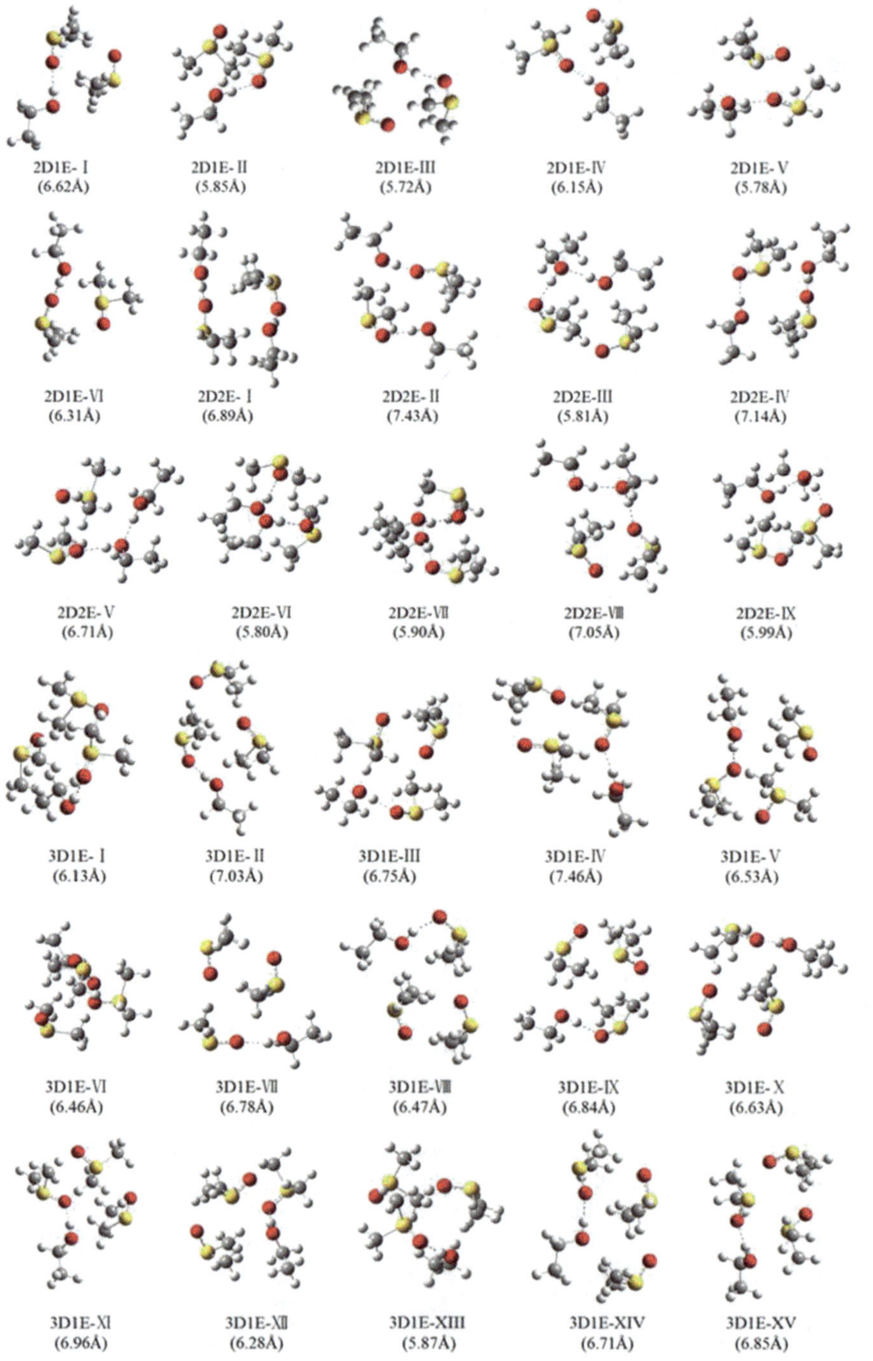

Fig.4.36 Optimized structures of $(DMSO)_m(C_2H_5OH)_n$ ($m = 2$, $n = 1$–2; $m = 3$, $n = 1$) molecular clusters along with its cluster radii

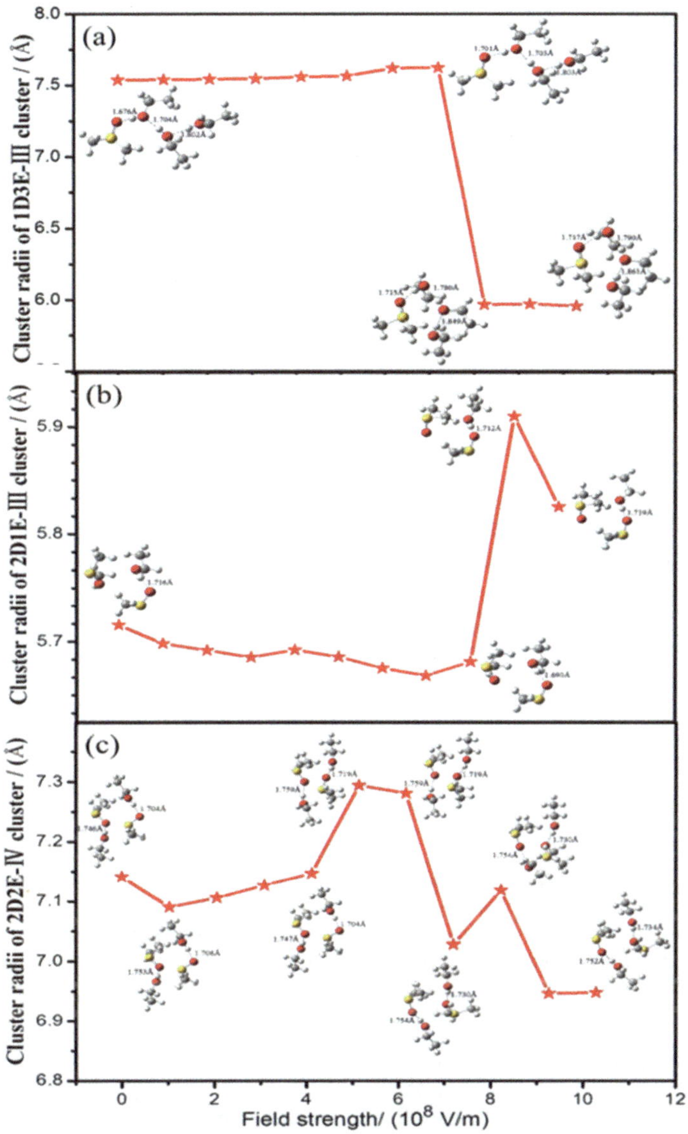

Fig. 4.37 Structural evolution and cluster radii of the binary molecular clustersalong with the the hydrogen bond length as a function of electric field strengths, (**a**) 1D3E-III, (**b**) 2D1E-III, and (**c**) 2D2E-IV

4.4 Summary

In summary, we introduced some special phenomena of interaction between microwave and materials, which are not consistent with the traditional microwave thermal theory. Therefore, it is necessary to study the interaction between microwave and materials at the molecular level to elucidate these special phenomena. We used MD method which is an important and effective tool at molecular level to simulate molecular dynamics under microwave field, and obtained a series of findings: (1) Compared to that of no microwave fields, the isotropic spatial collision probability of water molecules increases by 3% to 9% within the range of θ from 75° to 105° and the range of φ from 165° to 195° with the weak microwaves. (2) Microwave field reduces the number of water molecules in the water shell of ions, weakens the hydrogen bond strength and increases the conductivity of sodium chloride. (3) The simulation at molecular level also shows that some binary mixtures have special permittivity. (4) Furthermore, some specific clusters are found by the DFT calculation of binary clusters that affect the physical properties of the mixture, which provides a reasonable explanation for the special permittivity of the mixture.

These findings make people have a deeper understanding of the interaction between microwave and materials, and play an important role in guiding the application of microwave energy, especially microwave heating. This effectively solves one of the challenges for microwave energy applications, and provides foundations for large-scale applications of microwave energy.

References

1. Huang KM, Liu YQ, Tang JX, Zhao J, Wang BYA. Study of athermal action of electromagnetic waves on chemical reaction. Chem J Chin Univ. 1996;17:764–8.
2. Zuo F, Badev A, Saunier S, Goeuriot D, Heuguet R, Marinel S. Microwave versus conventional sintering: estimate of the apparent activation energy for densification of α-alumina and zinc oxide. J Eur Ceram Soc. 2014;34:3103–10.
3. Mazo P, Rios L, Estenoz D, Sponton M. Self-esterification of partially maleated castor oil using conventional and microwave heating. Chem Eng J. 2012;185-186:347–51.
4. Lahijani P, Zainal ZA, Mohamed AR, Mohammadi M. Microwave-enhanced CO2 gasification of oil palm shell char. Bioresour Technol. 2014;158:193–200.
5. Huang KM, Yang XQ, Hua W, Jia GZ, Yang LJ. Experimental evidence of a microwave non-thermal effect in electrolyte aqueous solutions. New J Chem. 2009;33:1486.
6. Gou DZ, Huang KM, Liu Y, Shi HX. Influence of weak microwaves on spatial collision and energy distribution of water molecules. Chem Phys. 2021;540:110977.
7. Hess B, Kutzner C, VanDer Spoel D, Lindahl E. GROMACS 4: algorithms for highly efficient, load-balanced, and scalable molecular simulation. J Chem Theory Comput. 2008;4:435.
8. Jorgensen WL, Chandrasekhar J, Madura JD, Impey RW, Klein ML. Comparison of simple potential functions for simulating liquid water. J Chem Phys. 1983;79:926.
9. Berendsen HJ, Postma JPM, Postma PM, van Gunsteren WF, DiNola A, Haak J. Molecular dynamics with coupling to an external bath. J Chem Phys. 1984;81:3684.
10. Essmann U, Perera L, Berkowitz ML, Darden T, Lee H, Pedersen LG. A smooth particle mesh Ewald method. J Chem Phys. 1995;103:8577.

11. Astrakas LG, Gousias C, Tzaphlidou M. Structural destabilization of chignolin under the influence of oscillating electric fields. J Appl Phys. 2012;111:074702.
12. Hess B. GROMACS 4: algorithms for highly efficient, load-balanced, and scalable molecular simulation. J Chem Theory Comput. 2008;4:116.
13. Van Gunsteren W, Berendsen H. A leap-frog algorithm for stochastic dynamics. Mol Simul. 1988;1:173.
14. Guardia E, Marti J, Padro J, Saiz L, Komolkin A. Dynamics in hydrogen bonded liquids: water and alcohols. J Mol Liq. 2002;96:3.
15. Elola MD, Ladanyi BM. Intermolecular polarizability dynamics of aqueous formamide liquid mixtures studied by molecular dynamics simulations. J Chem Phys. 2006;125:184506.
16. Lee H-S, Tuckerman ME. Dynamical properties of liquid water from ab initio molecular dynamics performed in the complete basis set limit. J Chem Phys. 2007;126:164501.
17. Chen C, Li WZ, Song YC, Yang J. Hydrogen bonding analysis of glycerol aqueous solutions: a molecular dynamics simulation study. J Mol Liq. 2009;146:23.
18. Zhang J, Dolg M. Global optimization of clusters of rigid molecules using the artificial bee colony algorithm. Phys Chem Chem Phys. 2015;17:24173.
19. Frisch MJ, Trucks GW, et al. Gaussian 03 (revision E.01). Wallingford, CT: Gaussian, Inc; 2004.
20. Lu T, Chen F. Multiwfn: a multifunctional wavefunction analyzer. J Comput Chem. 2012;33:580.
21. Humphrey W, Dalke A, Schulten K, Mol J. VMD: visual molecular dynamics. Graphics. 1996;14:33.
22. Zhao DG, Truhlar DG. The M06 suite of density functionals for main group thermochemistry, thermochemical kinetics, noncovalent interactions, excited states, and transition elements: two new functionals and systematic testing of four M06-class functionals and 12 other functionals. Theor Chem Acc. 2008;120:215.
23. Rai D, Kulkarni AD, Gejji SP, et al. Exploring electric field induced structural evolution of water clusters, (H2O)n [n = 9-20]: density functional approach. J Chem Phys. 2013;138:044304.
24. Rai D, Kulkarni AD, Gejji SP, et al. Methanol clusters (CH3OH)n, n = 3-6 in external electric fields: density functional theory approach. J Chem Phys. 2011;135:024307.

Chapter 5
Industrial Application

Chapters 2–4 mainly introduced some theories and models of microwave chemistry. From the above theoretical basis, some practical examples, especially in industrial applications, are needed to illustrate the use of these theories.

Here, we introduce two typical industrial applications of microwave chemistry theory. This chapter mainly introduces how microwaves are applied in these two industrial production lines from the aspects of theory and models.

5.1 Production Line of Microwave Drying Lignite

5.1.1 Introduction

Lignite is the lowest level of mineral coal. It has a high moisture content, low calorific value, easy weathering, and can spontaneously combust. These properties lead to a high transportation cost of a unit of energy, which is not conducive to long-distance transportation and storage. The direct combustion of lignite has a low thermal efficiency, and the emission of greenhouse gases is very large, so it is difficult to develop and utilize lignite on a large scale. In addition, as a raw material, lignite's conversion and utilization are limited. Lignite liquefaction, dry distillation, and gasification are needed to reduce the water content in coal to below 10%. It is difficult to reach to the quality of various user requirements if the processing of lignite is not upgraded. The processing of lignite upgrading becomes the key to efficient development and utilization of lignite. The improvement in the lignite quality process mentioned here refers to the changes in the composition and structure of coal during lignite dehydration, molding, and thermal decomposition processes. Lignite moisture is decreased and the calorific value is significantly increased after dehydration and quality improvement. This can not only prevent coal from spontaneous combustion but is also convenient to use in power generation, gas generation, and chemical industry.

K. Huang et al., *Dynamics in Microwave Chemistry*,
https://doi.org/10.1007/978-981-15-9655-1_5

Lignite is formed in a short period from coal, and the pores are extremely developed. These pores are long and narrow fractures with one end closed and the other end open. The surface and interior are full of organic matters and minerals, which have a strong ability to absorb moisture. Because of the property of these pores, lignite is a porous solid material with different pore size distributions. On the other hand, the lignite pore surface contains strong hydrophilic-rich oxygen containing functional groups, so the water content is higher than that of bituminous coal and anthracite coal. During the traditional drying process, the free water is first evaporated and the volume of lignite is reduced. The pore size decreases and the capillary action is enhanced, and so the evaporation rate of water is gradually reduced. The dewatering process is carried out in a much smaller pore structure. Water is formed by hydrogen bonding to the pore surface and the oxygen-containing functional groups, and then, these clusters are connected to form a monolayer or multilayer of water molecules. Due to the existence of hydrogen bonding, this water is more difficult to evaporate than that in the capillary pores. This is a fatal problem in the deep drying of lignite, which makes it impossible to promote the development of industrial microwave drying any further.

To solve this problem, Huang's group studied the influence of microwave radiation on the configuration of the intermolecular hydrogen bond in coal by using electrodynamics and MD simulations. Huang's group found that the hydrogen bond configuration clearly changes under high-power-density conditions (the field intensity is greater than 10^3 V/m), resulting in an easy removal of intermolecular water from the coal. The drying efficiency is greatly improved under this condition. Huang's group also found that drying lignite by microwave heating has a good energy-saving effect because of the high efficiency of microwave heating.

5.1.2 *Effective Permittivity Under Different Sizes and Temperatures of Lignite*

In the application of microwave drying, the dielectric property is key to investigate the interaction between microwaves and the media, and in the design of relevant devices. The introduction of dielectric properties was given in Chap. 2.

The complex permittivity of lignite at different temperatures and particle sizes are shown in Figs. 5.1 and 5.2. When the temperature is constant, and the average particle diameter is decreasing, the real part of the permittivity decreases, while the imaginary part increases. When the average particle diameter is a constant and the temperature decreases, the real part of the permittivity decreases, while the imaginary part increases.

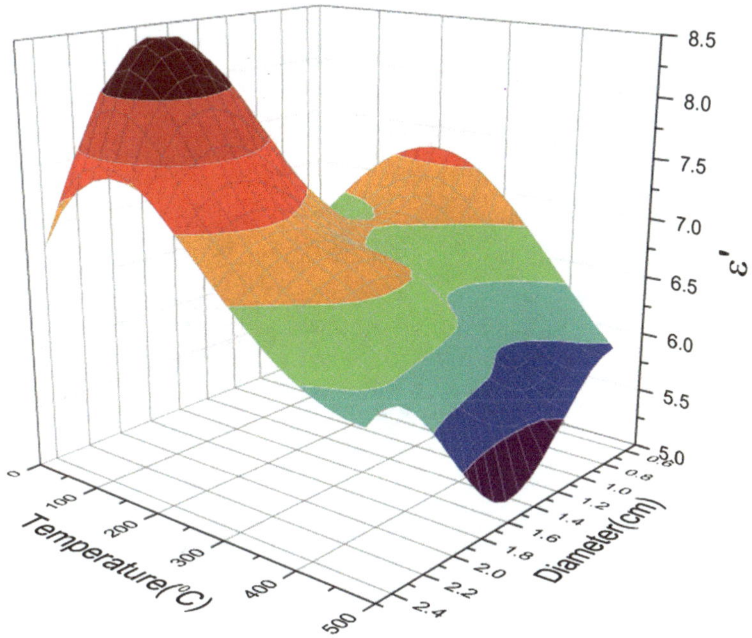

Fig. 5.1 Relation between the real part of the dielectric coefficient and the temperature and average particle diameter

5.1.3 Modeling of the Microwave Drying Process

From Sect. 5.1.1, it is known that microwave heating has some advantages in lignite treatment, but because of lignite's low burning point, uniform heating is very important, otherwise it is easy for lignite to spontaneously combust and lead to accidents during the drying process. A similar event occurred in the microwave drying production line of "Nanjing Sanle Microwave Technology Co. Ltd." Therefore, the microwave heating uniformity of the production line must be very high.

Because of the large scale of the production line (a single cavity with 3 m × 3 m × 1.7 m), ordinary commercial software and a single computer cannot perform the calculation (for example, the model has 69.2 million degrees of freedom in the multiphysics field calculation software), and there is no effective method to design and optimize the cavity structure. To solve the problem of uneven heating, Huang's group used the multiphysics field collaborative computing platform to design and optimize the structure of the production line cavity at "Nanjing Sanle Microwave Technology Co. Ltd," as shown in Fig. 5.3. The principle of the structure is that microwaves are generated from the feed port and enter the cavity through the horn antenna. Microwave distribution in the cavity is uniform by the juxtaposition of the feed port. Moreover, the size of the cavity affects the microwave distribution,

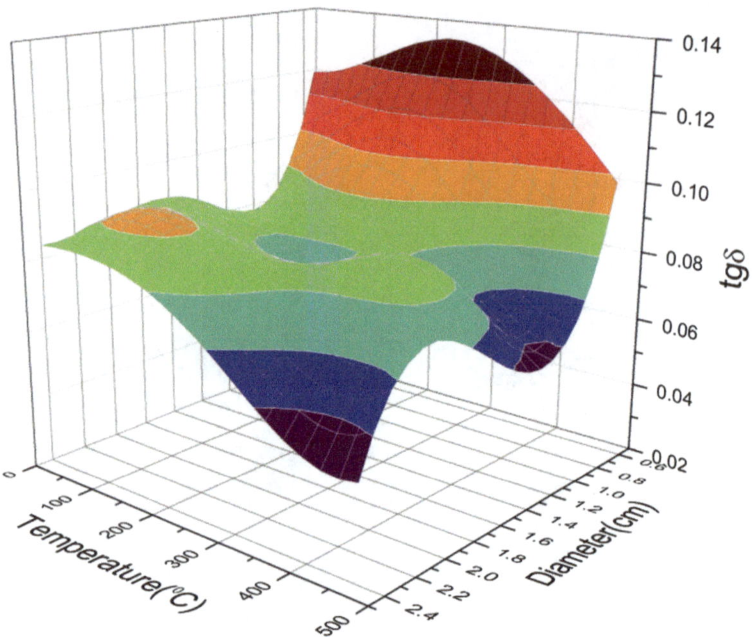

Fig. 5.2 Relation between the loss tangent and temperature and the average particle diameter

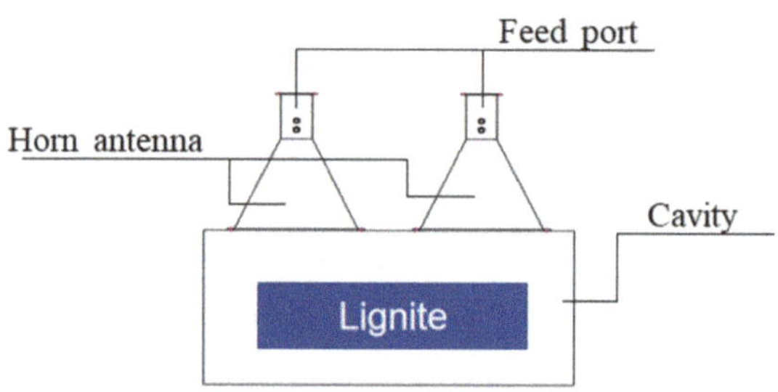

Fig. 5.3 Cavity and feed port structure

which affects the drying effect on the lignite, so Huang's group optimized the cavity size.

The temperature distribution is shown in Fig. 5.4a. It is evident that there are some points where the temperature is above 240 °C or below 60 °C. From the calculations, the temperature difference between the highest and lowest temperatures of the single cavity is 120.0 °C, and the average temperature is 62.45 °C. The temperature distribution before optimization is therefore uneven. After optimization, the cavity height is 1700 mm and the feed port size is 2400 mm × 1000 mm. The temperature distribution is shown in Fig. 5.4b, and the temperature distribution range is about 65–120 °C. The temperature difference between the highest and lowest

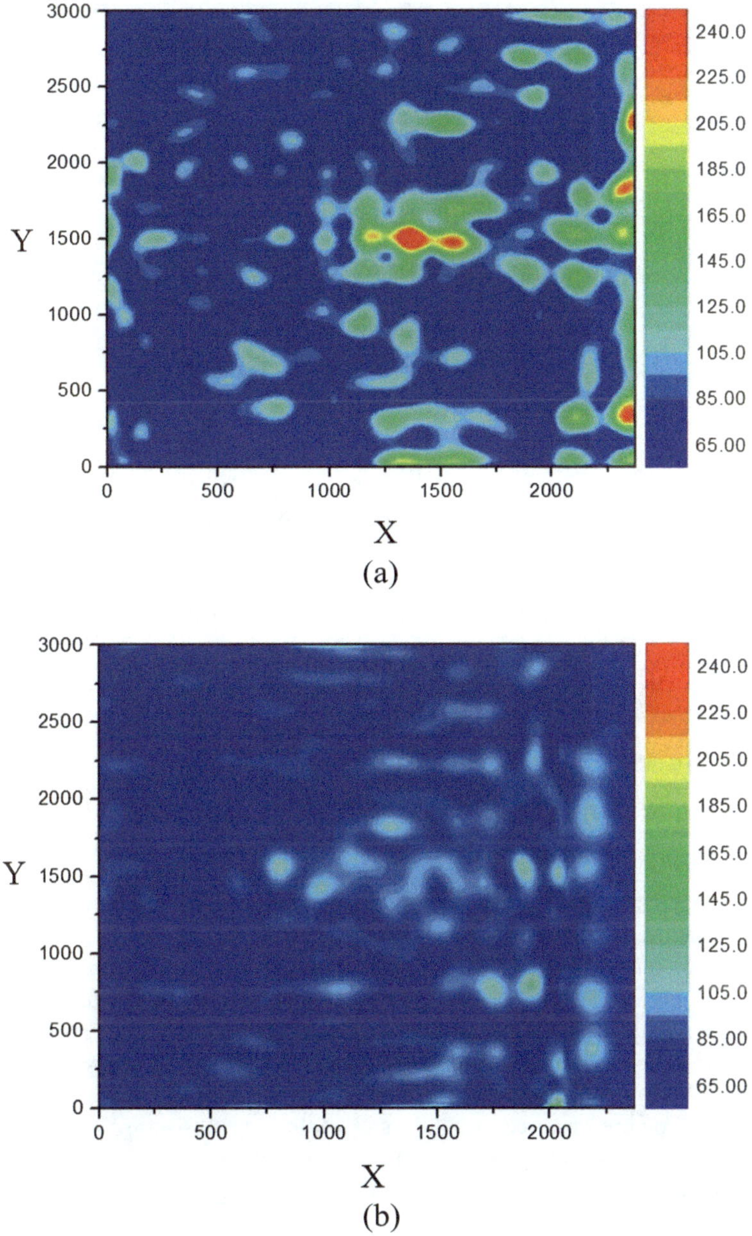

Fig. 5.4 Comparison of the temperature field distribution in a single cavity before and after optimization. (**a**) Before optimization and (**b**) After optimization

Fig. 5.5 Large-scale industrial production line

temperatures of a single cavity is 49.37 °C, and the average temperature is 67.83 °C, which is more uniform than that before optimization. Therefore, after optimization, the model has achieved the purpose of uniform heating and improved heating efficiency.

The current large-scale industrial production line and the microwave drying of the lignite production line are composed of several microwave sources, as shown in Fig. 5.5.

5.2 Production Line of Nickel Ore Smelting

5.2.1 Introduction

With the increasing investment in Chinese nickel laterite, Chinese high-quality laterite nickel ore resources have been significantly reduced. The laterite nickel ore contains less than 2% nickel and is cheap, but the cost of smelting by traditional processes is expensive. The effective development of low-grade laterite nickel ore has become a realistic problem. Laterite nickel oxide comes from oxidized sand and its domestic use is mainly from Indonesia and the Philippine, where it is the main raw material for nickel pig iron smelting and for producing stainless steel. High-grade nickel-iron ore resources are monopolized by Japanese companies, and China can only import low-grade nickel placers with a content of 0.9–1.1%.

At present, there are two main processes for smelting laterite nickel ore: a sinter furnace and a rotary kiln furnace. A sinter furnace first dehydrates and blocks the laterite in the sintering machine to form nickel sinter with a certain lumpiness and strength, and then smelts it in a furnace. In the mature technology of a sintering furnace, 1 ton of 10% NiFe consumes about 800 kg of pulverized coal, 600 kg of lump coal or coke, and 4800 kwh power to obtain a yield of about 92% for Ni. Smelting energy consumption is high and the economic benefit is low. Compared to traditional smelting, microwave smelting has some advantageous properties, such as fast heating speed, strong penetrating ability, and sensitive reaction. This method of smelting heats the interior and exterior together through the internal material to produce the heat source. In this way, the smelting time is greatly shortened, the smelting efficiency is high, and the product yield is improved. The following two sections will introduce the principle of microwave smelting, model design, and its application in a production line.

5.2.2 *Effective Permittivity Under Different Temperatures of Nickel Ore*

Due to the microwave heating processes temperature is very high, the nickel ore's dielectric properties under different temperature is measured by the equipment given in Chap. 2.

The effective permittivity of the nickle ore under different temperatures are shown in Fig. 5.6. The complex permittivity of laterite ore was measured using a designed on-line complex permittivity measurement system. In the experiment, it was found that the absorption characteristics of the laterite ore were lower than 1200 °C. When the temperature was above 1200 °C, the heating rate was faster when the microwave was heated, and the absorbing characteristics changed obviously.

5.2.3 *Modeling for the Microwave Process*

The concept of the process line as shown in Fig. 5.7, the equipment comprises a preheating section, a microwave heating chamber, an energy-fed waveguide group, and a microwave feeding port, with a microwave leakage suppression device installed at the end of the microwave heating cavity.

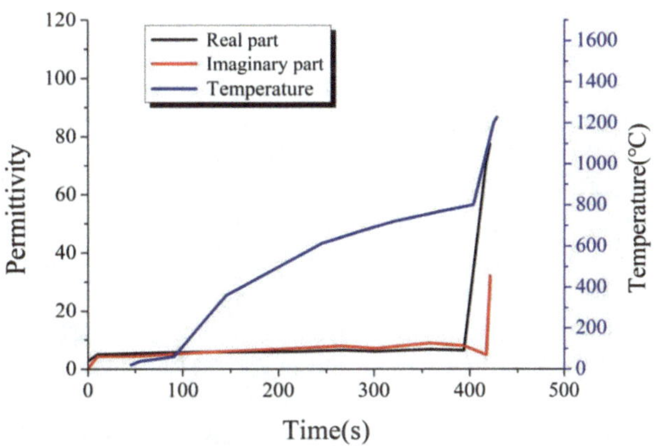

Fig. 5.6 The effective permittivity of the nickle ore under different temperatures

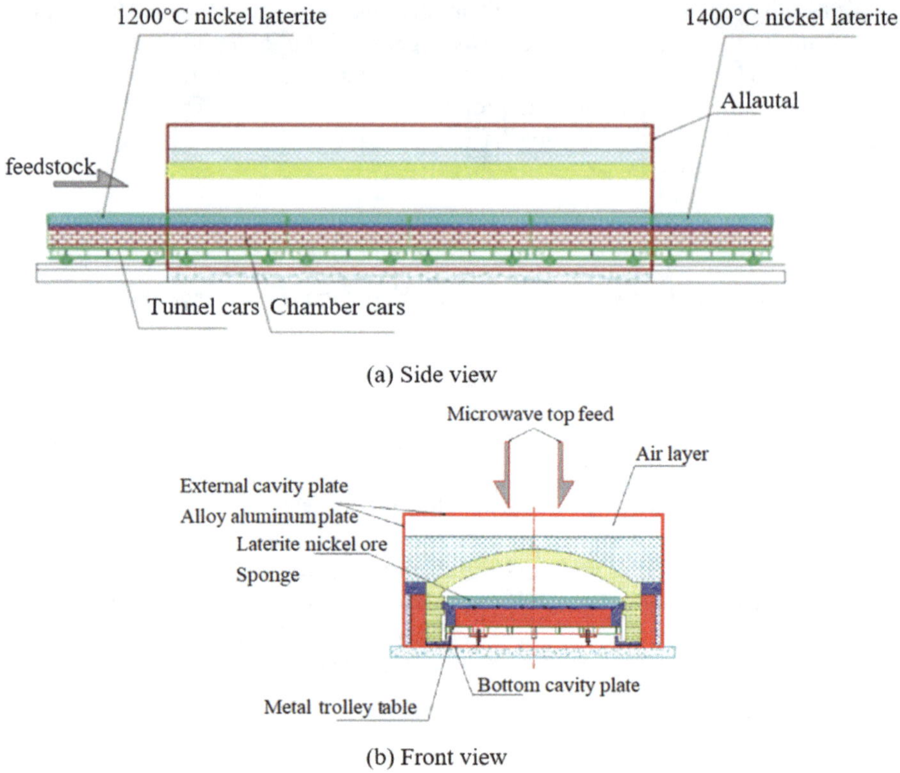

Fig. 5.7 Structure of the continuous-tunnel kiln-type nickel microwave smelting equipment

The microwave heating chamber is fed from the top. It is arranged by the adjacent vertical feed port; 50 horn-shaped power feeder ports are orthogonally arranged on top of each cavity, and each power feeder port is connected to a microwave source. In this way, the power density can be enhanced and the power output area can be more concentrated.

Except for the port arrangement, the port size, cavity size, and load position also affect the heating uniformity, and therefore, the productivity. The optimization objective is to obtain uniform and efficient heating. The horn aperture is 400 mm × 200 mm, the height is 250 mm, the cavity size is 10,000 mm × 5800 mm × 2880 mm, the load height is 200 mm, and the load is positioned 825 mm from the bottom of the cavity. The temperature distribution is shown in Fig. 5.8, where it is clear that although there are several hotspots in the processed area, they do not affect the overall temperature uniformity distribution because the temperature difference is small. Therefore, by this design, a more uniform temperature distribution can be obtained.

Based on the smelting equipment model presented in the previous section, a nickel ore production line was designed. The production line was built in Jiangsu

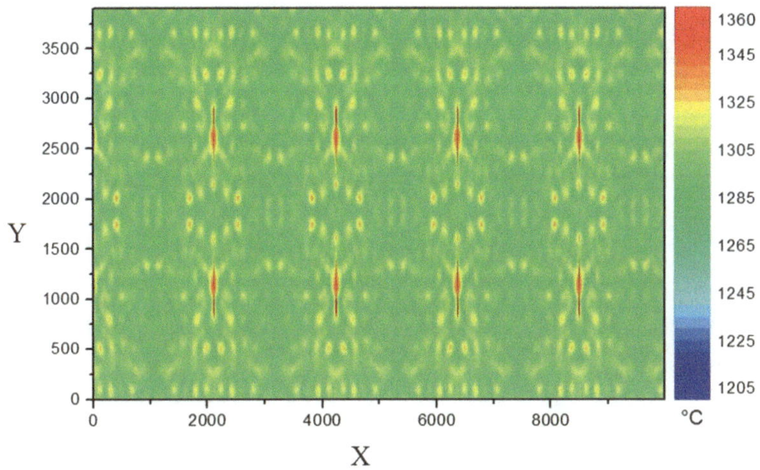

Fig. 5.8 Temperature distribution of the material

Fig. 5.9 Photograph of the Jiangsu Dafeng Port line

Dafeng Port, as shown in Fig. 5.9, and is currently in trial operation. The production line uses 50 microwave sources of 915 MHz, and each of them has a 75 kW power and a total power of more than 3 MW. The trial operation is in good condition at present. In the actual production, 1 ton of 10% NiFe consumes about 1500 kg pulverized coal and 1000 kWh electricity. Compared to the traditional method, the yield of Ni is 91% and the comprehensive energy consumption is reduced by 35%.

5.3 Summary

"Dynamics in Microwave Chemistry" is a new field that correlates the theory with the practice of microwave technology. This chapter described how the theoretical knowledge of microwave chemistry can be applied to industrial production through two industrial examples. It provides the necessary theoretical and technical guidance for the production rate of some large industries. With time, "Dynamics in Microwave Chemistry" will be increasingly widely used around the world.